To Lesley
& Helen —
two wonderful friends who
have gotten me through
many trials of life.
Love you both,
Pim

Shattered Minds

SHATTERED MINDS

How the Pentagon
Fails Our Troops
with Faulty Helmets

ROBERT H. BAUMAN
and DINA RASOR

Foreword by Perry Jefferies

Potomac Books

An imprint of the University of Nebraska Press

All rights reserved. Potomac Books is an imprint of the
University of Nebraska Press.
Manufactured in the United States of America.

Photos by Robert H. Bauman

Library of Congress Cataloging-in-Publication Data
Names: Bauman, Robert (Robert H.), 1941– author. |
Rasor, Dina, 1956– author.
Title: Shattered minds: how the Pentagon failed our troops
with faulty helmets / Robert H. Bauman and Dina Rasor;
foreword by Perry Jefferies.
Other titles: How the Pentagon failed our troops with faulty
helmets Description: Lincoln: Potomac Books, an imprint
of the University of Nebraska Press, [2019] |
Includes bibliographical references and index.
Identifiers: LCCN 2018028080
ISBN 9781640120365 (cloth: alk. paper)
ISBN 9781640121652 (epub)
ISBN 9781640121669 (mobi)
ISBN 9781640121676 (pdf)
Subjects: LCSH: Helmets—United States—History—21st
century. | United States—Armed Forces—Equipment—
Quality control. | Head—Wounds and injuries—United
States—Prevention. | United States. Department of Defense—
Procurement—Corrupt practices. | United States—Armed
Forces—Procurement—Corrupt practices.
Classification: LCC UC503 .B38 2019 | DDC 623.4/41—dc23 LC
record available at https://lccn.loc.gov/2018028080

Set in Sabon Next LT Pro by E. Cuddy.

Contents

Foreword

PERRY JEFFERIES

The grifters, con men, and thieves have accompanied every army ever put together into every foreign land and probably half the battles.

I am sure the Romans had their profiteering class, shorting the legions of their due paid in salt. European kings played many a game to finance their wars while enriching themselves. In America during the Revolution, Thomas Paine wrote about Silas Deane and Robert Morris, self-dealing war matériel in and out of Congress: "Is it right that Mr. Deane, a servant of Congress, should sit as a Member of that House, when his own conduct was before the House of judgment? Certainly not. But the *interest* of Mr. Deane has sat there in the person of his partner, Mr. Robert Morris, who, at the same time that he represented this State [Pennsylvania], represented likewise the partnership in trade."[1] Before the Deane affair was ended, Paine, whose pamphlets helped inspire the revolution, was beaten in the street; Deane and his partner Morris were discredited; and the president of the congress, Henry Laurens, resigned. According to Dean's biographer, Craig Nelson, this would "mark the beginnings of feral party politics in American government."[2]

In the Civil War, merchants and suppliers, known as sutlers and victuallers, followed the army, often selling back goods embezzled from supply rooms and warehouses. Government contracts were awarded under shady circumstances and fulfilled under even shadier ones. When the uniforms supplied to the Union Army by Brooks Brothers began to disintegrate because of the poor quality of the wool, the New York state legislature had to step in and replace them. A kickback here, a bottle of brandy there, to grease the permitting process, and one could get rich even in the midst of war. Millionaires in New

York City multiplied a hundredfold during the Civil War. This kind of graft is nothing new under the sun.

Smedley Butler, at the time the highest ranking and most decorated marine in U.S. history, declared, "War is a racket" in the 1930s.[3] He may be considered to have displayed as much courage after his service as he toured the nation and lectured diverse audiences, "trying to educate the soldiers out of the sucker class."[4] He wasn't the only military leader to do so, including one-time general of the Army and president Dwight D. Eisenhower, who worried that the perceived need for permanent war preparations would drain the resources of America and allow defense contractors to exert "the acquisition of unwarranted influence."[5] This concern was important then and is even more so now when it is in the interest of multinational companies to advocate for military solutions to world problems, solutions using their machines and technologies. If you've ever seen an advertisement for an F-35 fighter plane on television, ask yourself, who watching this show can buy that plane?

America, as in so many other fields, developed a high-tech solution to the issue of war profiteering—but not the one we wanted. While the Pentagon's books are still so poorly kept they cannot be audited, a 1982 law turned military logistics into civilian contracting, the first LOGCAP (Logistics Civil Augmentation Program) contract was born, and the coffers were open. A revolving door was installed between government and military service and the contract companies doing business with them.

Post 9/11 Bunnatine Greenhouse, a chief contracting office for the U.S. Army Corps of Engineers, tried to enforce the law as it applied to contractors writing contracts to themselves, but she was forced out of the office and the U.S. government and Iraq's doors were opened to KBR, Haliburton, DynCorps, and others like them. Once again old men were going to get rich on the backs and blood of America's youth, for these things indeed often come at a great cost to the younger and more vulnerable soldiers. While I served in Iraq, the ugly truth of profiteering and war stood out to me. When our soldiers had to exist on two packaged meals and two bottles of water a day, it stood out. When our troops walking on checkpoints had to use duct tape to hold their disintegrating boots together, it stood out. When my friends had to guard convoys delivering generators to contractor camps and our troops had none, it stood out. I was mad and wanted to do something about it.

So in January 2008, I found myself sitting next to Dina Rasor and Bob Bauman in Senate hearings. To our right was Stuart Bowen, the special inspector general for Iraq reconstruction, who had discovered and reported on billions of dollars of waste, which at that time had resulted in the recovery of over $17 million in funds, later reported to recover over $200 million.[6] I'd met Dina and Bob through my association with the Iraq and Afghanistan Veterans of America and shared my stories and those of some friends with them. Some of those stories ended up in Dina's second major book, *Betraying Our Troops*. Dina and her coauthor, Bob Bauman (former investigator for the Department of Defense's Defense Criminal Investigative Service), rooted out fraud, waste, and abuse of government systems. Dating back to Dina's (and any woman's) first ride in an M-1 tank in 1981, they have held fast and charted a deliberate path to the truth. Her first book, *The Pentagon Underground*, was an eye-opener for America, acquainting us with the $600 toilet seat and the $7,000 coffee maker. That book helped inform me as a soldier that not everything we were sold was gold. I was surprised I was able to buy the book in an Army Exchange bookstore. Dina and Bob provide this information as a service to the nation. Perhaps most importantly it is a service that impacts the foot soldier, the infantryman, and the tanker barely keeping her or his head above water while the grifters drive on gold-plated lanes.

Sadly, Stuart Bowen later ran afoul of ethics rules, and was fired from the Texas Health and Human Services Commission for unauthorized moonlighting in support of the Iraqi government.

Bob and Dina, however, remained true to their mission. They dug in, chasing the accounts shared with them by trusting soldiers and marines while rejecting the tales spun by senior military leaders and contractors. When the Army finally issued its recall of over one hundred thousand helmets that put our troops at greater risk for what we know as "the signature injury of the Iraq war"—traumatic brain injury—Dina and Bob were there to report. And to care for the troops.

The book you hold in your hand is a powerful tool in the fight for the welfare of U.S. troops. If waging that battle means taking on the leadership of the Pentagon, the professionals in the contract world, and all the grifters conning our nation, so be it.

Thank you, Bob and Dina. Thanks for staying the course.

Notes

1. Silas Deane, *The Deane Papers* (New York: New York Historical Society, 1889), 21:269.

2. Craig Nelson, *Thomas Paine: Enlightenment, Revolution, and the Birth of Modern Nations* (New York: Penguin, 2007), 133.

3. Smedley D. Butler, *War Is a Racket* (New York: Roundtable Press, 1935).

4. Jules Archer, *The Plot to Seize the White House* (New York: Hawthorn Books, 1973), 23.

5. President Dwight D. Eisenhower, farewell address, January 17, 1961, box 38, Speech Series, Papers of Dwight D. Eisenhower as President, 1953–61, Eisenhower Library, National Archives and Records Administration, Abilene, KS.

6. *Joint Hearing before the Federal Financial Management, Government Information, Federal Services, and International Security Subcommittee and the Oversight of Government Management, the Federal Workforce, and the District of Columbia Subcommittee of the Committee of Homeland Security and Governmental Affairs United States Senate*, Rep. No. Hrg. 110-437 (Washington DC: U.S. Government Printing Office, 2008), 28–29.

Introduction

Among surviving soldiers wounded in combat in Iraq and Afghanistan, TBI appears to account for a larger proportion of casualties than it has in other recent wars. . . . By effectively shielding the wearer from bullets and shrapnel, the protective gear has improved overall survival rates, and Kevlar helmets have reduced the frequency of penetrating head injuries. However, the helmets cannot completely protect the face, head and neck, nor do they prevent the kind of closed brain injuries often produced by blasts.

—New England Journal of Medicine, May 19, 2005

Rockets and mortars rained down on Lance Corporal Justin Meaders, U.S. Marine Corps (USMC), and his combat unit, the Seventh Engineer Support Battalion (ESB), as they fought in the streets of Fallujah. It was November 2004, and the Seventh ESB was part of the push to take the city of Fallujah from insurgents, a joint American, Iraqi, and British offensive code-named Operation Al-Fajr. As a combat engineer, part of Justin's job was demolition, including blowing up weapon caches and houses where weapons were stored and conducting counter-improvised-explosive (IED) sweeps, also known as C-IED in military lingo. The fighting units depended on him and his men to clear paths for them and find weapon caches and "drop houses" that were suspected of holding weapons or enemy fighters.[1]

During a weapons cache sweep at the outskirts of Fallujah, Justin's marine unit, in support of another marine unit, carefully approached a building that they suspected contained weapons. Two of the marines entered the building. The lead marine inadvertently walked over a concealed IED, which exploded, instantly killing him. His backup, who was about sixty-five feet behind him, was hit hard by the explosion. When he was pulled out of the building, his face was swollen, and fragment wounds peppered his body. The overpressure from the blast had caused all the capillaries in his face to explode, but his helmet had stayed on. He had altered the inside of his helmet, fitting it with commercial foam pads as an upgrade from the old leather-band suspension system with a half-inch air gap between the helmet shell and his head.[2]

The Marine Corps had not provided the men with a padded suspension system for their helmets. It was not standard issue. Army soldiers had them in their new helmets, but the marines did not.[3] The helmet pads, however, were vital for survival in combat, especially with the proliferation of IEDs, but marines who wanted pads for their helmet had to obtain them outside the normal supply chain. The wounded marine was medically evacuated to an Army hospital in Germany and survived. Justin firmly believed the foam pads greatly enhanced his chance of survival.[4]

The combat helmet is the singular piece of equipment adopted to protect the head of soldiers and marines in combat. However, the helmets since World War I have been limited to protection from the impact of bullets and shell fragments, known as ballistics. Until recently, protection from nonballistic blunt and blast-wave impact resulting from ordnance like IEDs was largely not considered in the design of the combat helmet. As a result blast-wave and blunt-force impact was a major cause of mild traumatic brain injury (mTBI) for the troops in the Afghanistan and Iraq conflicts. Unlike blunt-force impact, the blast wave is a pulse of highly compressed air that spreads outward from the center of an explosion slamming into a person's brain nearby at a speed faster than sound.

Significantly, a nine-month Rand Corporation study published in 2008 estimated that 320,000 soldiers had suffered mild to severe traumatic brain injuries during deployment in the Iraq and Afghanistan conflicts, mostly from the effects of exposure to blasts from IEDs.[5]

In 2011 the Department of Defense (DOD) reported that 361,000 soldiers had suffered brain injuries since 2000 even though many experts believe the number is over 400,000.[6] Joe Rosen, a professor of surgery at Dartmouth Medical School, believes that as much as "eighty percent of the injuries coming off the battlefield are blast-induced."[7] Despite the military's recent ambition to upgrade a helmet that would provide protection against blast impact, an earlier effort to upgrade could have saved a considerable number of soldiers from traumatic brain injuries.

The DOD defines traumatic brain injury (TBI) as a traumatically induced structural injury or physiological disruption of brain function as a result of an external force impact to the head.[8] The medical world defines TBI as a "nondegenerative, noncongenital insult to the brain from an external mechanical force, possibly leading to permanent or temporary impairment of cognitive, physical, and psychosocial functions, with an associated diminished or altered state of consciousness."[9] TBI can be classified based on its severity—mild, moderate, or severe—but most attention has been focused on mTBI, leading to its designation as "the signature injury of the war."[10]

TBI has been an injury in all wars but historically associated only with apparent penetrating head wounds from firearms and other projectiles. Closed head injuries that resulted in mild TBI were invisible injuries, sometimes misrepresented as more psychiatric than physical. The long-held belief was that if the head wasn't bleeding, it was fine, an attitude that lasted well into more recent conflicts.

During World War I, the term "shell shock" was used to describe symptoms, now associated with TBI, of soldiers exposed to exploding ordnance. The similarity between the shell shock of World War I and the current injuries of TBI was noted by Edgar Jones, among others, in the *American Journal of Psychiatry* in 2007: "We note that the U.S. military currently committed to serious fighting in Iraq and Afghanistan faces a situation similar to that of the British Army engaged in the Somme offensive of July 1916. Both campaigns have developed into wars of attrition in which head wounds and concussion are common battle injuries. The high casualties of the Somme battle brought the issue of shell shock to the fore when, as traumatic brain injury has done today, it caught the popular imagination and the attention of the media."[11]

In World War II, soldiers exposed to explosions showed the symptoms now associated with TBI. These symptoms were then termed "combat neurosis," "exhaustion," "post-trauma concussion state," or "combat fatigue."

During the Vietnam War, such symptoms were called "Vietnam Syndrome," as if closed head injuries were exclusive to that war. In 1980 the American Psychiatric Association termed symptoms from combat exposure as post-traumatic stress disorder (PTSD). PTSD remained a catchall diagnosis into the early stages of the conflicts in Afghanistan and Iraq to describe unexplainable psychiatric and other emotional symptoms of soldiers exposed to combat. It wasn't until later in those conflicts that medical officials started recognizing many symptoms, previously considered PTSD, as mild TBI—the result of closed head injuries from blunt-force impact and, more significantly, exposure to blast waves.

TBI that is mostly caused by exposure to IED detonations has accounted for a higher percentage of casualties in the Afghanistan and Iraq conflicts than in any previous wars. For instance, in the Vietnam War, TBI injuries represented only a small percentage of casualties. The reason for the difference was the introduction of the Kevlar helmet, which although it provided a higher degree of ballistic protection, did not protect the soldier or marine from TBI injury resulting from exposure to a blast.

The combat helmet was in use as far back as 1015 BC. Composed of various types of metals and other materials, it was so heavy and cumbersome that it limited mobility on the battlefield, but it was effective against the weapons of that time. Helmets were used throughout the Middle Ages, but once firearms became the weapon of choice on the battlefield in the fifteenth century, the use of the helmet declined, and by World War I, it had completely disappeared from use.

However, during World War I the loss of thousands of soldiers due to head injuries led to a comeback of the helmet. The French developed the first metal helmet, which resembled a soup bowl, called the Adrian, named after the French Army general August-Louis Adrian. The British soon followed suit with the production of the steel Brodie helmet, designed and patented in 1915 by self-styled engineer John Leopold Brodie. Soon after the introduction of the Brodie helmet, metallurgist Sir Robert Hadfield redesigned it using harder steel that offered better protection; the result was the Hadfield helmet.

The United States had no helmets in its inventory when it entered the war in 1917. Instead, it purchased Brodie helmets from the British to equip its troops, renaming the helmet the MK I. By 1918 the United States was producing and shipping its own helmets, a modified version of the Brodie helmet, to use during the war. The idea behind those steel helmets was to protect soldiers from shells lobbed overhead into the trenches.

The Germans in World War I came up with their own steel helmet design that was far superior to the Adrian and the Brodie in its design and protective capabilities due to its use of harder steel. Called the *Stahlhelm*, it was designed by Dr. Friedrich Schwerd of the Technical Institute of Hannover after he had studied head wounds suffered during trench warfare.

The *Stahlhelm* was based generally on a fifteenth-century helmet that provided good protection for the head and neck. It was the only helmet during World War I that included a padded suspension system consisting of a headband with three segmented leather pouches holding padding materials that provided a comfortable fit. The helmet was fielded to the German troops in February 1916 at Verdun and resulted in a dramatic decrease in serious head injuries. The German World War I design became the most successful combat helmet in both that war and later in World War II.

Since World War I, the American combat helmet has gone through a number of iterations. Experimental efforts to come up with a better design continued during and after World War I to develop a distinctly American helmet. One design, like the German helmet, included a suspension system with three cushioned pads attached at separate points around the band to provide stability and comfort. However, the three-pad suspension system in tests did not provide the comfort sought, and in 1934 it was determined to be unacceptable. Thus in 1940 a new steel helmet design called the M-I (known as the "steel pot") utilized a webbed suspension system, also called a sling suspension system. It consisted of an adjustable leather band fastened around the head, along with supporting straps, separating the helmet shell from the skull. The M-I was a big leap forward for combat helmets and featured a lightweight liner.

From the end of World War II through the Vietnam War, there were a number of design improvements to the steel helmet that mostly affected the inside liner to tailor it for various military occupations such as ground troops, artillery, airborne, and signal corps. In 1958 F. J. Lewis of the Naval Medical Field Research Laboratory issued a helmet design report that "indicated the great importance of the suspension system." He stated, "The suspension may be the deciding factor regarding level of protection, compatibility under various environmental conditions, and acceptance of a helmet design by the individual wearer and his unit" and called for the use of multiple pads over the cranial area as the practical suspension system.[12]

In 1968 an alternate helmet design to the M-I helmet known as the Hayes-Stewart helmet, after Office of the Surgeon General Brigadier General Hayes and Edgewood Arsenal's George Stewart, was developed for use in the Vietnam War. The design resembled a Roman helmet and incorporated a removable cushion-type suspension system consisting of five sponge pads. Tests were conducted to determine stability, compatibility with equipment, evaluation of human factors engineering, such as comfort and safety, soldiers' opinion of comfort, and acceptability to soldiers. It was also found to be more stable and preferred by soldiers over the M-I helmet.

In 1969 modifications were made to the pads, resulting in a sandwich-type construction with polyethylene and polyurethane foam. However, by 1972, after many tests with both a padded and a sling system, the padded suspension system was considered unsatisfactory even though it was more comfortable and equal to the MI with respect to stability, preference, and effect on performance. After rejecting the padded suspension system, the Army helmet bureaucracy decided to improve the sling suspension system for the M-I helmet.

It wasn't until 1983 that the Army began fielding a new and revolutionary combat helmet made of Kevlar that would provide better ballistic protection. The Kevlar shell was molded out of multiple layers of woven Kevlar panels, giving the shell enough density for maximum ballistic protection. However, it took the Army's time-consuming bureaucracy eleven years to develop, test, and produce the Kevlar helmet. Called the Personnel Armor System Ground Troops (known as the PASGT) helmet, it would become the standard-issue helmet for ground troops and paratroopers for the next twenty years. The Kevlar PASGT helmet design was nicknamed the "Fritz" after its resemblance to the successful German helmet. Although the PASGT helmet would provide better ballistic protection, it differed from the German helmet with its padded suspension system. Instead, the PASGT helmet included the standard sling suspension system used in the old M-1, which would not provide important protection from nonballistic and blast injuries.

Despite the PASGT helmet's increased ballistic protection, Army helmet designers had not considered protection from blunt-force or blast impact along with comfort and stability. Helmets for paratroopers were only slightly modified—a little padding for the nape of the neck. It did little good, and incidents of head injuries to paratroopers continued.

The Army helmet bureaucracy appeared to be satisfied with the poorly designed PASGT helmet during the 1990s, but soldiers were not. In a June 2015 online report on combat helmet protection, Christian Beekman notes, "Only 30% of PASGT users were satisfied with their helmet's maintainability, and 15% were satisfied with its fit. Less than 10% of PASGT users were satisfied with their helmet's comfort, weight, and overall impression."[13]

Not satisfied with the PASGT helmet, the Army's Special Operations Command (SOC), on its own, redesigned the PASGT helmet to give SOC soldiers better head protection, comfort, and utility; the new helmet was fielded in 2001. Unlike the regular military, the SOC discovered the tremendous benefits of including a seven-pad suspension system on the inside of the helmet that provided effective protection from blunt impact and blast injuries. Later, research confirmed that a helmet with a padded suspension system decreased blast overpressure to the head by as much as 86 percent, while a padless helmet with an inside gap such as the PASGT and any other helmet with a sling suspension system increased blast overpressure to the head.[14] Adding foam pads to the inside of the helmet was a big step forward for the soc in limiting the effects of exposure to blast and TBI.

It became clear during the conflicts in Iraq and Afghanistan that a newly designed combat helmet was needed to protect the troops from TBI, especially from the effects of exposure to blast but also from blunt impact. But the regular military helmet bureaucracies stubbornly continued to resist change to the PASGT helmet for the ground troops until the Army finally took a cue from Special Forces and developed the new Advanced Combat Helmet, modeled after the soc's helmet, which included a padded suspension system. However, the Marine Corps defied change with its new Lightweight Helmet (LWH), sticking to the old padless suspension system. Resistance to change by these bureaucracies is not a new problem. It has been around since before the Civil War. Conflicts of goals and mission between the military's operating forces in combat and the bureaucracies in Washington have always existed, many times at the expense of the protection of the troops.

This resistance has also fostered many whistleblowers from both inside and outside the military, all trying to effect change for better protection for the troops. However, the military bureaucracies have fought against these intrusions with a "not invented here" attitude. Senator Charles Grassley testified in a congressional hearing on government whistleblower retaliation, stating, "In 33 years, under both Republicans and Democrats, I've found the problem the same. Whatever bureaucracy you're talking about, whistleblowers are about as welcome in a bureaucracy as skunks at a picnic."[15]

Fraud, abuse, ethical violations, or bureaucratic malpractice have been problems exposed on many occasions. Whether it is those who report contractor fraud or ethics violations on military projects, or those both inside and outside the DOD who try to expose abuses or malpractice on military programs, whistleblowers are normally hit with retaliation such as demotion or termination or, at the very least, rejection of their concerns. Some within government have their security clearances revoked or are even charged with a criminal violation. Most whistleblowers are people who are only trying to do what they perceive as the right thing—to correct injustices or other violations.

According to a 2011 national business ethics survey, retaliations against whistleblowers have risen to a new high.[16] Most of these whistleblowers went no further than lodging their complaints through internal channels because they feared retaliation. Others took it further by filing a federal action against a company through the *qui tam* provision of the Federal False Claims Act.

Abraham Lincoln instituted the *qui tam* law in the 1860s as a means to prevent defense contractors from selling bad equipment and sick horses to the military during the Civil War. Lincoln in effect deputized average citizens by allowing them to sue on behalf of the government if they had direct knowledge of fraud, and the citizens (called relators) shared in a percentage of what was recovered by the government. If the government intervened in the case, the reward was usually lower than if the relators took the case through the courts themselves. However, if the government intervened, there was a much higher chance of a monetary recovery. The *qui tam* provision law as enacted under Lincoln fell into disuse because subsequent modifications made it almost impossible to use. It was revived in 1986 because of a rise in defense contractor fraud during the increased defense spending in the 1980s. The revised law also has some of the strongest whistleblower legal protections in the federal government.

Taking on the DOD bureaucracy is not easy because it is a unique entity—a self-sustaining empire with its own culture and language and even its own procurement rules. The DOD could be considered almost a fourth branch of government, almost untouchable by the other three, especially Congress, which goes out of its way not to mess with the ever-increasing defense budget. This special status of the DOD is why it is so problematic for outsiders to make improvements or changes within DOD agencies. Without transparency it is very difficult to determine accountability for procurement decisions made in the DOD's Byzantine bureaucracy.

We began our research for this book with one issue in mind: the story of whistleblowers Jeff Kenner and Tammy Elshaug and their battle with Sioux Manufacturing over shorting the weave on Kevlar panels used to mold the shell of the PASGT helmet. While doing a series of initial interviews and other research, we stumbled upon activist Dr. Robert Meaders, nicknamed Doc Bob, and his nonprofit organization Operation Helmet. We found that his campaign for the use of a padded suspension system for combat helmets to protect soldiers from blunt impact and exposure to blast was vitally important for soldiers' health. Also, we found his story to be compelling for his epic battles with the military helmet bureaucracies over something as important as head protection for the troops. Thus this book focuses on the helmet as a whole—the weakened Kevlar shell with Jeff and Tammy's story and the inside suspension system of the helmet with Doc Bob's activism to provide a padded suspension system for all helmets worn by the troops to protect them from nonballistic and blast injuries in a war in which TBI was on the rise.

However, as we acquired more information, it became evident that the common denominator linking Jeff and Tammy's story with Doc Bob's was the military bureaucracy's "not invented here" attitude, which was bent on maintaining, at all costs, its projects, position, careers, jobs, and a competitive edge in the fight for funding. The resultant stance turned out to be the real center of the story. This bureaucracy engaged in serious malpractice that ignored or prevented improvements to the helmet that might have saved many soldiers from TBI. This bureaucracy failed by being uncooperative in a criminal investigation, resisting all efforts by Doc Bob to upgrade its helmet suspension system and undermining a helmet pad vendor's effort to provide a quality product to the military. While the military bureaucracy was resisting Doc Bob's efforts to provide more head protection for the troops, more and more veterans were developing serious brain injury from exposure to blast. Along with a dramatic rise in veteran TBI, mostly caused by exposure to the effects of blast, came a dramatic rise in suicides triggered by the deteriorating effects of brain injury.

Abbreviations

ACH	Advanced Combat Helmet
ADS	Atlantic Diving Systems
APL	Applied Physics Laboratory
ARL	Army Research Laboratory
BCT	brigade combat team
BFD	backface deformation
BIRI	Brain Injury Research Institute
BLSS	Ballistic Liner and Suspension System
CGF	Casgues Gallet Francaise
CID	Criminal Investigations Division
CNRM	Center for Neuroscience and Regenerative Medicine
CREW	Citizens for Responsibility and Ethics in Washington
CTE	chronic traumatic encephalopathy
DARPA	Defense Advanced Research Projects Agency
DCIS	Defense Criminal Investigative Service
DCMA	Defense Contract Management Agency
DIC	digital image correlation
DLA	Defense Logistics Agency
DOD	Department of Defense
DOJ	Department of Justice
DOJ-OIG	Department of Justice-Office of Inspector General
DOT&E	director, Operational Test and Evaluation
DVBIC	Defense and Veterans Brain Injury Center
ECH	Enhanced Combat Helmet
ESB	Engineer Support Battalion
FAR	Federal Acquisition Regulations
GAO	Government Accountability Office

GMMC	Guard Material Management Center
GSA	General Services Administration
HAHO	high altitude–high opening
HALO	high altitude–low opening
HMSS	Helmet-Mounted Sensor System
IBA	Interceptor Body Armor
IED	improvised explosive device
IHSS	Improved Helmet Suspension System
JTAPIC	Joint Trauma Analysis and Prevention of Injury in Combat Program
LWH	Lightweight Helmet
MICH	Modular Integrated Communications Helmet
MRAP	Mine-Resistant Ambush-Protected
MSA	Mine Services Appliances
mTBI	mild traumatic brain injury
NCO	noncommissioned officer
NIB	National Industries for the Blind
NGREA	National Guard Reserve Equipment Account
NPR	National Public Radio
NRL	Naval Research Laboratory
NVGS	night vision goggles
OBL	Oregon Ballistics Lab
OSI	Office of Special Investigations
PARC	principal assistant responsible for contracting
PCO	program contracting officer
PEO	Program Executive Office
PTSD	post-traumatic stress disorder
PUU	polyurethane urea
QAR	quality assurance representative
RFI	request for information
RFP	request for proposal
SFTT	Stand for the Troops
SMART-TE	Suspension Material Analysis and Retention Technology-Test and Evaluation
SMC	Sioux Manufacturing Corporation
SOC	Special Operations Command

TBI	traumatic brain injury
TES	traumatic encephalopathy syndrome
PASGT	Personnel Armor System Ground Troops
UHMWP	ultra-high molecular weight polyethylene
USAASC	U.S. Army Acquisition Service Center
USAARL	U.S. Army Aeromedical Research Lab
USSSOC	U.S. Special Operations Command
USPFO-CIF	U.S. Property Fiscal Officer, Central Issue Facility

Authors' Note

In the case of titles, including those of the Defense Department and other government entities cited in this book, we follow *The Chicago Manual of Style*, the publisher's preference, which calls for capitalizing such titles only when they precede a personal name or when capitalization is used in a direct quote.

Shattered Minds

CHAPTER I

Hard-Headed Marines

I was serving in Iraq as a civilian bomb disposal contractor. On May 3rd, 2006 my convoy was hit by an IED. The bomb went off next to my truck injuring myself and one of my security team members and unfortunately killing our driver. I sustained injuries to my face requiring facial reconstruction, multiple shrapnel wounds, and tendon loss in my right arm. My doctors were extremely surprised that I had not sustained any brain damage. I was wearing a helmet outfitted with this [Oregon Aero pad] kit. I just wanted to write a quick note saying "thank you." On behalf of myself, my family, and my friends: Thank You!!! Please keep up the good work.

—Email to Dr. Robert Meaders, Operation Helmet

Justin Meaders knew from a young age that he wanted to spend his life in the military, and he aspired to be a marine. His role model was his grandfather, Dr. Robert Meaders, known to everyone as Doc Bob. He was a career Navy doctor. Justin started high school at one of the lowest-performing schools in Houston. He did poorly in that school environment, barely making passing grades. To pass all a student had to do was show up. Justin found there were more parole officers in the hallway than teachers. He wanted something more structured and with more discipline.

His grandfather suggested the Marine Military Academy in Harlingen, Texas. It was the only military academy (high school) he was familiar with, and it had a good reputation. The academy contributed many enlisted men to the Marine Corps, much like West Point and the Naval Academy produce officers for the Army and the Navy, respectively. Justin liked the discipline and the sense of order the school offered, so he enrolled. He did very well, and his grades dramatically improved. During his senior year, school officials told him he would make a wonderful marine, and they really wanted to have him in the corps. It was Justin's dream come true.

In August 2003 at age nineteen, Justin enlisted in the Marines. Boyish looking, soft-spoken, with a low-key demeanor, Justin did not seem to fit the persona of a tough marine who would delight in large and noisy demolitions. After attending infantry combat training at Camp Pendleton, California, he chose Combat Engineer School at Camp Lejeune, North Carolina. He wanted to use his hands and knew that combat engineers have a wide variety of jobs other than simply running a machine gun. They conduct demolition missions and handle construction assignments such as building survivability positions and bridges. After completing the engineering school, he was assigned to the Seventh Engineer Support Battalion based at Camp Pendleton.

In June 2004, right after Justin was settled with his new unit, they were ordered to prepare for deployment to Iraq. "We started doing predeployment build-ups and did a Joint Task Force Six counter drug operation with the Border Patrol and Army Corps of Engineers along the border of Mexico," Justin recalled. "That's where I found out about foam pad upgrades for the PASGT helmet."[1]

Justin's first experience with wearing the PASGT combat helmet occurred while he was in U. S. Marine Corps boot camp in San Diego, California. "It was horrible, especially the leather band liner," he recalled. "It slipped around my head, was sweaty, and would stick to my head, especially if I had to paint my face up. Also, it would wobble around on my head and whenever I had NVGS [night vision goggles] mounted on the helmet, it would slip down on the front of my head because of the extra weight."[2]

Under the broiling desert sun of summer along the border with Mexico, Justin and his unit were working hard during a counter drug operation. "We were building something, survivability positions, simple, rough things like that," he explained, "and I noticed an Army sergeant, first class, wearing a PASGT helmet that seemed to be secured to his head, while working, unlike everyone else's helmet." Curious,

Justin approached the Army sergeant and asked him about his helmet and how it stayed on his head so well.[3]

The sergeant, just back from Iraq, gathered the troops around and told them, "Look at my helmet. It's got an upgrade to it, and if you're smart you'll get one for yourself. It might save your life, like mine." The sergeant asked Justin, "Would you like to try it on?" "Sure," Justin replied. He put on the helmet, which, unlike his own PASGT, had impact-protecting pads and a four-point suspension system. It felt secure and very comfortable on his head, especially with the better four-point retention system instead of a single chinstrap. He was so impressed that he asked where the sergeant had found it. "I bought it online from a company called Oregon Aero," the sergeant revealed to Justin. Justin decided, right then, that he had to get a pad kit for his helmet before deploying to Iraq.[4]

The initial appeal of the pads was solely for comfort and stability. At that time comfort was a priority for a new marine. Justin was not aware of TBI and the safety benefits of the pads. These young marines felt immortal and didn't talk about being injured or killed.

Shortly after his return to Camp Pendleton, Justin used a friend's computer to email Oregon Aero and inquire about ordering pad kits for his whole unit, hoping for a discount. After three or four tries, he didn't receive a response and couldn't figure out why. Undeterred but running out of time, Justin called his grandfather to discuss the pads and look into a way of contacting the company and ordering them at a fair price. The Oregon Aero BLSS (ballistic liner and suspension system) kit contained seven highly engineered foam pads as well as a four-point restraint system and nape pad, providing stability, along with blast and impact protection, just what the Marines needed.[5]

A native Texan Dr. Robert Meaders has always been a man of action, one who went to great lengths and sacrifice to help those in need. A retired Navy flight surgeon who, among many distinguished assignments in an adventurous twenty-one-year career, had jumped from planes on search-and-rescue missions, picked up astronauts after splash down, run a provincial hospital in the Mekong Delta of Vietnam during the war there, and provided medical training in the

Ethiopia highlands to Ethiopian doctors while overcoming intermittent assaults from local bandits. In Vietnam he witnessed firsthand the cost of war, having to conduct facial reconstructions for soldiers who had suffered significant facial and head injuries. He also treated brain injuries and witnessed their devastating effects.

Unfortunately his adventurous career with the Navy came with a physical cost. During a training exercise to rehearse rescue missions while stationed in Guam, Doc Bob made a parachute jump from an altitude of about five hundred feet when his parachute lines fouled, causing him to hit the ground at thirty miles an hour. He broke three vertebrae in his back and ended up with lifelong back problems. He was medically retired from the military in 1979 because his bad back made him wheelchair bound. Yet it was against his nature to settle for a life in a wheelchair.

Determined not to remain in a wheelchair, Doc Bob took a year off, immersed himself in intensive physical therapy, and slowly recovered the ability to walk. Not content with "retirement," he joined the International Eye Foundation in Bethesda, Maryland, and was sent to live and work with primitive tribes in Africa tending to their particular health problems. During a six-year period, his work took him to many parts of Africa while he acted as a consultant to the World Health Organization for its prevention of blindness program.

After Doc Bob returned from Africa, he entered private practice in Mesa, Arizona, as a retinal specialist. Still not ready to be constricted to a comfortable office, he flew his plane to the periphery of Indian reservations, where he helped institute a clinic to treat the local citizens in Casa Grande and especially the Yuma Indians, who were suffering a very high incidence of diabetic retinopathy. However, his practice of medicine finally came to an end in 1994, when he suffered a ruptured intestine, exacerbated by the many parasitic diseases he had contracted during his work in Africa.

Doc Bob now lives in Montgomery, Texas, in a gated community of landscaped homes surrounded by two private golf courses about fifty miles north of Houston. He remains nonetheless an unpretentious man, who likes to explain that he was born in 1934, "on the kitchen table of [his] house" in Glen Rose, Texas, a small town about sixty miles southwest of Dallas/Ft. Worth, "delivered by an eye, ear, nose, and throat doctor who happened to be passing by."[6]

With his grandson's request for helmet pads in hand, Doc Bob decided to thoroughly research the use of the padding before he bought them. "I spoke to Navy explosive research doctors I knew who were doing blast research and asked them what they knew about helmet pads. They told me they buy that stuff from Oregon Aero and stick it in their helmets to do blast studies," he recalled.[7]

However, Doc Bob is not one to rely on one person's opinion. He wanted more verification. He contacted international companies specializing in blast protection research in France, Israel, Netherlands, and Canada, and they all confirmed the use of Oregon Aero foam pads for impact effect and testing. He found studies from a Canadian group conducting blast research and head protection on the internet. These studies were done in conjunction with a U.S. Navy panel set up to examine head injuries. Their general opinion was that shock waves did create a threat for brain injuries, and Oregon Aero was the only pad maker whose product was recommended to prevent or mitigate this type of injury.

Aris Makris, a scientist with the Ottawa, Canada, research group for Allen-Vanguard, Med-Eng Systems, specialized in blast protection and injury research. His research was primarily of use in mine removal efforts (called "demining") worldwide. He conducted a study revealing what he called the "scoop effect" of the PASGT helmet, which with only the sling (padless) suspension system, can enhance a blast wave impacting the head because of the "air gap" created by a standoff, the separation between the helmet shell and the head. This impact can lead to head concussive trauma. According to Makris, "foam pads can take care of this issue and will generally be a positive development in filling the air gap between the head and shell." However, Makris warned that "the foam, if too solid/stiff can serve to better transfer the [blast] wave into the head."[8]

Searching the internet Doc Bob found a November 2000 article published by the U.S. Army's Natick Soldier's Center (Natick) claiming that a soldier could survive a shot to the head wearing the newly designed Modular Integrated Communications Helmet (MICH), used

by Special Forces, and "get back into the fight" because of "the innovative seven-pad suspension system."[9] Impressed with the possibilities of marines using helmet pads, he called George Schultheiss, the
helmet manager at Natick. Schultheiss praised Oregon Aero pads,
saying they were the best available pads and the Army was using
them as the padded suspension system in its new Advanced Combat Helmet (ACH).

Schultheiss explained that at first only the Army's Special Forces
Command used Oregon Aero pads, but the Army was now buying
them by the hundreds of thousands to provide the ACH to its regular soldiers. However, Schultheiss pointed out that the Marine Corps
had not approved a padded suspension system for its new LWHS,
soon to be fielded to its troops. The LWH was being issued with the
old leather headband suspension system, the same suspension system used as far back as World War II and more recently in the PASGT
helmet. Schultheiss assured Doc Bob that the padded Oregon Aero
BLSS system was a worthwhile buy for his grandson.

It didn't make any sense to Doc Bob that the Marine Corps would
make the effort and spend the money to develop a new helmet and
ignore the protective value of a padded suspension system for its
troops. The PASGT helmet was engineered to protect against ballistics and provided only fair protection from blunt-force impact,
blast forces, and fragment impacts from IEDS. It depended on an
old 1935-era headband, or sling, suspension system to "float" the helmet over the head, maintaining helmet and head separation. And
now, Doc Bob thought, the Marines were still going to use the old
headband system in their new LWH and that would not protect
their troops any better than the old PASGT against nonballistic or
blast impact. He believed that a shock-absorbing pad suspension
system would be far superior, as the Army's MICH had proved, in
providing standoff protection from blunt-force impact and blast-
wave overpressure and fragments.

Maybe the Marine Corps would supply Oregon Aero pads to Justin and his unit to retrofit their PASGTS, Doc Bob naively thought.
He believed that the corps would embrace his request, as it would
add a level of protection the helmet lacked. He contacted officials of

the Marine Corps Combat Equipment Team in Quantico, Virginia, and asked them. Their answer angered him. They would not retrofit PASGT helmets with pads due to what they claimed were budget limitations.[10] A Marine Corps official said, "BLSS kits are not authorized or needed in the new LWH. We have not prohibited the use of BLSS kits to retrofit old helmets but do not see the need to buy these kits as we are meeting our needs with the new helmet."[11]

The Marine Corps is a parochial bureaucratic institution, so its response was not a surprise. Public bureaucracies, especially hardened military bureaucracies, operate within a set of predetermined rules, standards, policies, and decisions based on past experiences and the current politics within the bureaucracy. They rarely act contrary to their rules or policies, especially in response to what they consider an outside intrusion.

Almost thirty years earlier, the Marine Corps PASGT helmet bureaucracy had decided that its helmet would not include a padded suspension system. Its new LWH was likely based on the same material and same suspension system as the PASGT's. The design addressed weight and ballistic impact but apparently not comfort, stability, or nonballistic impact. On the other hand, the Army had based the design of its ACH on the MICH, which addressed nonballistic impact, comfort, and stability and included the padded suspension system. The Marine Corps's decision regarding what type of helmet marines would wear in combat was approved at the highest level of leadership and funded with taxpayer's dollars. However, standards were established and would not be changed by present-day reality or outside interference.

Doc Bob thought of writing his congressman but felt it would not solve the problem. Undeterred and driven by his desire to protect his grandson and his fellow marines, he decided to act directly. If he wanted Justin to have Oregon Aero pads, he was going to have to pay for it and send the pads directly to him. He also thought that the only fair thing to do was to buy the pads for any soldier or marine who requested them.

Completely convinced that Oregon Aero's product was first rate as protective equipment for his grandson's helmet, Doc Bob called the company. Buying them out of his own pocket at $118 per kit, Doc Bob ordered a dozen pad kits, the same pad kit used in the MICH by Special Forces, and sent them to Justin and enough for his rifle team. After wearing the helmets with the pads installed for a few days of training, an excited Justin called his grandfather. "Opa, these are wonderful," he said, "but we feel like we can't wear these unless we can get them for our whole company." Doc Bob swallowed hard and said, "Well, how many of those do we have to come up with?" "A hundred," Justin hesitantly replied.[12]

Doc Bob funded the initial $11,000 for Justin's company, but he soon realized much more needed to be done. Determined to solve the helmet problem for other marines, in addition to Justin's company, he decided to raise money for the helmet pads and retention systems by walking the neighborhood and explaining to his neighbors how important the pad kits would be to the marines in combat. His neighbors gladly contributed money, and he was able to raise enough money to buy not only the kits that Justin and the rest of his company needed before they deployed to Iraq but also enough to ensure a steady stream of them should it become necessary.

A month into his deployment in Iraq, Justin reported back to his grandfather that the pads worked very well—the marines wore their helmets all the time, even taking naps with them on. Also, his commanding officer, Maj. M. N. Hess, saw the value and the positive effects of the helmet pads for his troops during missions. Expressing his gratitude in a memorandum to Doc Bob, in October 2004 Hess formally thanked him for saving his marines over $10,000. Major Hess wrote that his marines wore Kevlar helmets for long periods of time to protect them from enemy fire. They found that Oregon Aero pads were noticeably more comfortable, which made their lives easier. He thanked Doc Bob for significantly improving "Marines' quality of life and their force protection posture."[13]

It wasn't long before marines from other units learned about the helmet pads. They saw Justin and his company of marines running across the sand without having to hold their helmets with one hand and their weapons with the other, able to fall to a prone firing position without having to slap their helmets back, not having to hold their helmets up to keep their eyes uncovered when wearing night-vision goggles, and strangely enough, sleeping using their helmets as a pillow. Not unexpectedly, they also wanted the pads.

Before long requests for pads started pouring in to Doc Bob. Word of the pads also spread to Army soldiers, and soon Doc Bob was receiving requests from regular Army units along with National Guard and U.S. Army Reserve units that had deployed to Iraq, still using the old padless PASGTs. Doc Bob knew that he had started something. Now he needed to continue raising money to fund more kits for all the troops who requested them for use in combat.

Eventually, on the advice of an attorney friend, Doc Bob started a nonprofit organization so that people's contributions would be tax deductible. He named it Operation Helmet, and Justin Meaders would be a co-founder. Next, he established the Operation Helmet website, which explained the safety aspects of using helmet pads in a "360-degree" war where the troops faced an enemy that used high explosives from all quarters, inflicting casualties from blast force and the large primary and secondary fragments that result. He pointed out that with a blast impact, a helmet without pads would "rock" on the head, making violent contact with the skull in an area about the size of a ball-peen hammer. More importantly, this type of impact, if severe enough, caused skull fractures, intracranial bleeding, and concussions, possibly resulting in death or disabling injuries.

The solution Doc Bob offered was to replace the unstable suspension system of the PASGT and the new Marine Corps LWH with a shock-absorbing pad suspension system featuring a four-point restraint system that provided more stability. When a soldier or marine sustained impact on the helmet, the padded suspension system absorbed the energy of the impact, dispersing it and preventing the helmet from contacting the head, while providing comfort and stability. Therefore, the upgrade cushioned the trooper's head from blast and fragment impact and thus could save lives and prevent disability or TBI.

CHAPTER 2

Twisted Logic

Within the Defense Department there was the top-down message that said, "Suck it up, we don't need to hear about this"—especially the MTBI, even though there were lots of people saying that this was a serious problem.

—Dr. James Kelly, director of the National Intrepid Center for TBI at Walter Reed National Military Medical Center in Bethesda, Md. to the Huffington Post

B lunt impact tests conducted by the Army's Aeromedical Research Lab (known as USAARL) proved that a foam-padded suspension system would provide a higher level of blast and blunt-impact protection for both soldiers and marines.[1] Also in 2003 the Marine Corps's own team at Natick conducted impact attenuation tests of its then new LWH without a padded suspension system with the old padless PASGT and the all-foam suspension system MICH. The results showed that the four-point retention padless suspension system of the LWH offered some improvement in impact protection compared with the PASGT but found the all foam-padded MICH to be "superior" for impact protection.[2] This result would prove problematic for Marine Corps helmet officials who had designed the LWH without a padded suspension system. Despite the results of the Natick tests and the fact that the Army already had included a padded suspension system with its new ACH, they ignored their own tests. Instead, they decided that it was unnecessary to include an energy-absorbing foam-padded suspension system and opted to continue exposing many of their marines in combat to potential TBI with their new LWH. It was a classic case of a Pentagon bureaucracy that would not or could not admit that another military service or outsiders would have a better and more innovative idea. This bias is known in the military bureaucracy as "not invented here" and has for years been a factor in buying planes, ships, and tanks. Now it was influencing decisions about the common marine helmet despite the threat to the Corps's own marines and their brains.

Doc Bob thought he was providing a service to the Marine Corps by alerting them to a way of mitigating brain injuries and that they would embrace his contribution. Instead, they ignored his plea to better protect their marines. To Doc Bob, the Marine Corps's stance was typical of its "tough, deal with it" attitude, as if marines were somehow impervious to TBI. He mused that it was like the 1943 battle of Tarawa in World War II, when the marines went ashore on the small island of Betio, on the western side of Tarawa, a place that was nothing but a big hunk of hard, jagged coral. "The guys aboard the Navy ship started taking their mattresses off their bunks, wrapping them up tightly, and giving them to the marines. Soon, as they got ashore with them, Lt. Col. Chester Puller, the storied marine officer, pulled up and said, 'what the hell is that you got on your bedroll?' 'It's a mattress sir,' was the response. Puller then had all the mattresses thrown into the bay. 'What the hell are you trying to do, make sissies out of my marines? Sleep on the coral!'"[3]

Despite the many marines desperately requesting pad kits from Operation Helmet for their helmets based on possible life-and-death consequences, Marine Corps officials claimed they were receiving "overwhelmingly positive feedback from its performance (the newly fielded LWH) are steadily meeting the demands of our warfighters."[4] They were perpetuating the typical Marine Corps cultural attitude. It made no sense to Doc Bob.

Continuing to ignore the Marine Corps's nonsensical stance was something Doc Bob could not tolerate; it was not in his makeup. He wanted a systemic change that helped all marines, not just the ones he sent pads to. He strongly believed the marines deserved better protection. He wanted to convince the Marine Corps that it should install pads in all its LWHs as the Army had done with its ACH. If the Marine Corps did that, then Doc Bob would have no further need for Operation Helmet, and he could just concentrate on playing golf.

Doc Bob's fund-raising efforts were covering the costs for the requests not only from marines but also from National Guard and U.S. Army Reserve units and Navy Seabees, who were using the old PASGT helmet without a padded suspension system. He started with the Marine Moms group in Houston, the Marine Parents, and the Marine Motorcycle Club, a group of men who were former snipers and reconnaissance marines. The Marine League was another organization that jumped in to help. These groups held poker runs, auctions, dances, and other activities, and the funds started building up. On one Saturday afternoon, they were able to raise $18,000. "We had an awful lot of help from people who felt a responsibility to show their support of the troops and were thinking about them. There was a lot of loyalty among ex-servicemen of any kind," Doc Bob recalled. He couldn't believe that a Boy Scout group was able to raise $1,500 by holding a pancake breakfast. It was that kind of grassroots effort that helped raise enough money to fund pad kit requests.

The pad kits were being provided to whole units, rifle teams, squads, and companies that were in or about to go into combat. Doc Bob would not send pads to marines or soldiers assigned to duty "inside the wire" (a term used to describe soldiers who are assigned to base duty in Iraq). He rarely received individual requests as marines always took care of their brothers-in-arms.

The documented impact tests showing the superior effects of padded helmet systems became important evidence as Doc Bob argued his case with the Marine Corps equipment bureaucracy. To add to those military-conducted tests, he decided to seek independent blunt-impact tests. Searching online for a reputable lab, he discovered Intertek, a multinational product-testing company located in Cortland, New York. Oregon Aero volunteered to arrange and pay for the tests of the Marine Corps's PASGT with and without pads at Intertek's testing facility.[5]

In early March 2005, Intertek test results showed that the PASGT with the BLSS kit installed significantly reduced blunt-impact force with an average of 79 g's (g represents a measurement of a gravitational force of the type of acceleration that is caused by the impact between two objects). This was far superior to the LWH without pads, which averaged a much worse 188 g's.[6] That high g-force number was considered a failure by Army standards. Doc Bob again appealed to the Marine Corps leadership to consider the benefits of using pads in the Corps's helmets. He emailed the Marine Corps program manager for individual combat equipment, a civilian employee of the Marine Corps, providing him with the test results and also forwarding a copy to the Marine Corps commandant.

About that time, the Marine Corps began fielding the LWH without a padded suspension system. Doc Bob contacted Marine Corps helmet officials, strongly urging them to change the interior of the LWH to include helmet pads instead of the old-fashioned leather sweatband and nylon-mesh skullcap.

After a couple of weeks with no reply, Doc Bob became increasingly upset with the Marine Corps's seemingly dismissive attitude and apparent disdain for outside interference. On March 21, 2005, he resent his email to the equipment program manager, but instead of replying to his emails, helmet officials attempted to co-op Doc Bob and Operation Helmet's effort to supply marines with pad kits with a so-called point paper justifying their decision to field the LWH without pads. This decision was approved up the Marine Corps chain of command, with a succession of officers buying into the seriously flawed justification that pads were not necessary for the LWH.[7]

In April 2005, Dan Fitzgerald, the Marine Corps's infantry combat equipment program manager, authored a point paper for the commanding general of the Marine Corps Systems Command titled "Lightweight Helmet (LWH) Sling Suspension vs. Ballistic Liner and Suspension System (BLSS)." Fitzgerald revealed that the point paper was initiated after a "growing amount of attention . . . generated by Dr. Bob Meaders, over using Oregon Aero BLSS pads for the LWH." He falsely accused Doc Bob of generating sales for Oregon Aero "in the name of providing BLSS kits to servicemen and women participating in OIF and OEF." Fitzgerald obviously perceived Doc Bob's initiative as a threat to the Marine Corps's decision making, noting that Doc Bob had claimed "that the BLSS [was] far superior to the LWH sling suspension system in terms of non-ballistic injury and should replace all LWH suspensions."[8]

However, the paper seemingly agreed that Oregon Aero's BLSS pads installed in the LWH did provide blunt impact protection, almost double the protection. Incredibly, it dismissed this finding, stating that the padded system did "not provide enough of an increase in non-ballistic protection to warrant its use in the LWH." Fitzgerald also erroneously claimed that the BLSS system pads were "highly absorbent" of sweat and other liquids.[9]

The point paper did not clarify what was not "enough of an increase in non-ballistic protection." Based on previous test results, the BLSS pads would have provided more than double the impact protection. Strangely, the paper dismissed the Intertek test results "based on the results of non-ballistic impact testing" that did not "translate into better ballistic protection." "Bizarre" was the only word Doc Bob could come up with to describe the Marine Corps's effort to tie the use of pads to ballistic protection over nonballistic protection.[10]

Also weighing in on the point paper was the Marine Corps's acting medical officer, G. R. Cox, who submitted a memorandum to the commanding general for the Systems Command, agreeing with Fitzgerald's position that "after reviewing currently available data, the LWH Sling Suspension [was] recommended over the BLSS for use by Marine Forces operating in combat environments."[11] The paper did not cite what "available data" he had reviewed. Given the fact that plenty of evidence existed disputing Cox's contention, he was either completely ignorant of the current research literature regarding the use of helmet pads or was just blindly following Marine Corps helmet bureaucracy dogma.

Doc Bob's anger grew. He felt this twisted logic made no sense. It was clear that Marine Corps helmet officials were in complete denial that helmet pads provided blast and blunt-force impact protection for their marines at a time when TBI was spiking in Iraq because of a huge increase in IED explosions. "Although the BLSS is effective in protecting the wearer from non-ballistic impacts such as riding in a pitching combat vehicle, falling debris, and falls," the medical officer wrote, "other considerations including comfort, fit, and performance when exposed to cold temperatures and moisture make the system unsuitable for operational use." Cox did not mention exposure to blast, and, again, he did not cite the factual basis for his statement that a padded suspension system was "unsuitable for operational use."[12]

Military institutions often have disdain for outsiders who intrude on their programs. In this case the Marine Corps's helmet bureaucracy was operating in a bubble. It was ignoring the fact that many of their marines were going to suffer TBI from nonballistic impact just to defend their decision to not use pads. Doc Bob noted that although the new LWH was eight ounces lighter and 30 percent more effective in ballistic protection than the older PASGT helmet, it failed the operational requirements for nonballistic protection. He believed that the helmet bureaucrats were resisting helmet pads because their use did not originate with them.[13]

Yet the point paper glossed over one important fact: most marine casualties in Iraq and Afghanistan resulted from repetitive blast-wave impact to the head acting as concussive forces, not just from crashes and falls. Fitzgerald neglected to consider flying gear, rifles, rocks, heated air, walls, wood, cargo, and other items becoming projectiles or fixed anvils transferring energy into the troops' heads during blasts and crashes. Casualties also occurred as marines worked on rooftops, doing war zone reconstruction, and were forced to jump out of the way of danger. All these events involved potential injury due to concussive forces. Also, when IEDs explode, or ambushes commence, U.S. troops have been trained to drive at high speeds using extreme, evasive tactical maneuvering to avoid rocket-propelled grenades (RPGS), ambush fire, or successive IED detonations.

The stakes in this debate were high for marines in combat. Their special operations counterparts had the benefit of the padded MICH, which had tested favorably by the services for both ballistic and nonballistic impact protection.

According to the Defense and Veterans Brain Injury Center (DVBIC) at the Walter Reed Army Medical Center, the Iraq and Afghanistan campaigns produced higher rates of TBI patients than previous conflicts.[14] Common sense recognizes that g-forces from IED blasts, close-support munitions, crashes, and sequential collisions in Iraq and Afghanistan will injure the brains of troops without padded protection at a higher rate than those of troops with the additional padded protection.

Doc Bob knew that head injuries and deaths in Iraq and Afghanistan were due more to nonballistic impact—blast and blunt-force trauma—than to ballistics. The Marine Corps's equipment bureaucracy was ignoring the real problem of head injuries caused by the ever-increasing IED blasts prevalent in Iraq. Even more ridiculous to him was its other reasons to justify not using pads. The Corps claimed that marines preferred the LWH with the sling suspension system to the BLSS pad kit because the pads were susceptible to trapping sand that caused abrasions. It had not indicated if this preference was that of marines surveyed in combat or those queried while on bases in the United States.

Highly skeptical that marines preferred the helmet without pads, Doc Bob was getting a different response from marines in the field that was exactly the opposite. In 2005 marines requesting pad upgrades commented that the BLSS pads did not absorb sweat or trap sand as the Marine Corps brass claimed. Instead, they said the outer material of the pads actually wicked sweat away from the wearer's skin.[15]

"The BLSS system is not an authorized modification to the LWH," the point paper predictably concluded in its recommendation. It also noted that the USMC should publish its official position regarding Doc Bob. This meant the Marine Corps leadership was not going to allow their marines to install pads in the LWH obtained privately through Operation Helmet. Knowing that he was dealing with an obstinate bureaucracy, Doc Bob resolved to continue sending marines the pads whenever they requested them. At the time the Army included a padded suspension system, using Oregon Aero pads for its newly fielded ACH. Apparently, the Army either did not share its information with the Marine Corps regarding the use of pads or the Marine Corps had dismissed the idea. "Nobody tells the marines what to do," Doc Bob thought.

Doc Bob plugged away at more research while sitting at his computer in his home office. As he gazed out at one of the fairways of the golf course adjacent to his home, he wished he were out there swinging a golf club instead of fighting the Marine Corps. He found that historically sling/strap suspension systems have worked well in maintaining head/helmet separation to prevent or mitigate behind-armor blunt trauma in a ballistic situation. However, nonballistic injury from impact and blast wave was not considered.

Despite the Army's 150 g impact limit requirement for its ACHS and MICHS, the Marine Corps held fast to its decision that the sling suspension system was enough (ballistic) protection without adding the nonballistic protection of shock-absorbing pads. Doc Bob felt this position was a direct challenge to the Army, Navy, and Special Forces' decision to utilize shock-absorbing pads to provide the best combination of ballistic and nonballistic protection for their troops.

To further refute logic spelled out in the point paper, Doc Bob cited a 2000 press release by the Army's Soldier Systems Center in Natick, Massachusetts, that revealed the MICH used by Special Forces soldiers actually improved ballistic protection, acting as a "shock absorber" for projectiles with a seven-pad suspension system from Oregon Aero with the "resilience of a wrestling mat." The use of the pads allowed "the wearer to stay conscious . . . and get back into the fight."[16]

Contrary to the Marine Corps's claim that Oregon Aero pads "were highly absorbent . . . of sweat and other liquids . . . and were susceptible to trapping sand causing abrasions," the Army's press release noted that the pads utilized "a black CoolMax cloth covering [that] wicks moisture away and helps the user stay cooler."[17] Doc Bob added that "the pads, coated with a semi-permeable membrane (think fish gills) demonstrated the ability to repel water, remain air-filled, and wick perspiration away from the wearer's head with Cool-Max coating."[18]

However, for Doc Bob this was not just a bureaucratic fight. The Marine Corps's intransigence was not good enough for his grandson and all the other marines fighting "a dedicated and devilish foe using roadside bombs, IEDS and other blast forces to inflict casualties," he wrote on his website.[19]

In November 2005 Doc Bob fired back with a rebuttal to the point paper, summarizing his arguments in a letter sent to Maj. Gen. William Catto, commanding general of the Marine Corps Systems Command. He argued that Oregon Aero's BLSS Kit provided superior nonballistic and blast effect protection compared to the sling suspension system. However, the author of the point paper said that the added protection from a padded suspension system did not provide "enough of an increase in non-ballistic protection to warrant its use in the LWH." Doc Bob wanted to know how much protection was "enough" for marines. "I want all I can get for my grandson and his comrades."[20]

Doc Bob also pointed out that some Marine Corps equipment officials questioned the BLSS system, asserting that it could cause "greater head trauma . . . given there is no standoff." However, he had explained to them that "the BLSS maintains standoff via a 'sandwich' pad, with a rapid-crush layer to allow helmet surface deflection in a ballistic strike and slow-crush layer to absorb and spread the injurious impact force of non-ballistic strikes. Doc Bob had concluded that those officials had "never dissected or tested the BLSS system, and rel[ied] on comments from Mr. Fitzgerald for their information."[21]

Despite the point paper's claim that "the sling suspension (a helmet without pads) meets or exceeds all requirements as described in the LWH ORD (Lightweight Helmet Operational Requirements Document)," Doc Bob countered in his letter to Major General Catto that had "it actually fail[ed] the non-ballistic protection test standards of other services and passe[d] only the lower standards of the ORD." He also argued that "it would make sense that a combination of the superior ballistic protection of the LWH combined with the superior non-ballistic impact protection of the BLSS Kit (or any other tested and proven shock-absorbing pad system) would provide the best available combination of protective factors in a potentially 'great helmet.' 'Perfect' should not become the enemy of 'great.'"[22]

Doc Bob made sure to let Catto know for the record that Operation Helmet had "no relationship or association with Oregon Aero" and was a volunteer organization with no administrative or employee expenses. He wrote, "100% of donations are utilized to send shock-absorbing pad systems to troops in the field. . . . Our sole aim is to help provide the best protection available to our troops in the field, with the help of families, friends and other sources desiring to support our troops in a meaningful fashion. We are a purely volunteer-run organization with no salaries or fees deducted from donations."[23]

With an increasing number of brain injuries and sound scientific testing showing the benefit of adding shock-absorbing pads for the LWH, Doc Bob believed the Marine Corps needed to revamp its official position on the use of helmet pads to provide additional protection from nonballistic impact. Whether it was the PASGT or the LWH, the Marines needed a padded suspension system in its helmets like that in the Army's helmet, and he was going to fight the narrow-minded Marine Corps bureaucracy to get it for the troops.

• • •

After finishing his first deployment to Iraq in March 2005, Justin let his grandfather know the padded helmets worked "amazingly well" and provided significant protection along with being very comfortable and stable on his head. The death of one of his fellow marines from an IED explosion had affected him deeply. The marine had been wearing a PASGT with Oregon Aero pads, but he was right on top of the IED when it exploded. Additionally, Justin knew a marine who, in the summer of 2006, was close to an IED explosion and suffered TBI. He was not in Justin's unit, but Justin knew he was wearing the new LWH without pads. With his grandfather's research, Justin was now even more thankful he had the pads for safety purposes.[24]

All through training, pre-deployment, and deployment in Iraq, Justin never removed or washed his pads. After returning from deployment, he sent the pads to his grandfather, who cut them up and found that they were like new on the inside. The outside was black with soil from the heat and the battlefield, but the foam was perfectly intact. Doc Bob was now more determined than ever to mainstream the helmet pads into the military. He was also extremely frustrated that he had to run a small nonprofit to get the troops what they needed while the Marine Corps and Army bureaucracies wasted billions of dollars and still did not adequately protect their troops.

CHAPTER 3

Moment of Truth

BISMARCK, N.D.—A former North Dakota college wrestling star killed while on duty in Iraq enlisted in the National Guard while in high school, following in his father's footsteps, military officials said.

Lance Koenig, 33, of Fargo, a former All-American wrestler for North Dakota State University, was killed by a bomb while on Guard patrol Sept. 22, Guard officials said.

Koenig enlisted in the Guard in 1988, as a high school student, and was assigned to Company B of the North Dakota Guard's 141st Engineer Combat Battalion, Guard officials said. His father, Robert, also was a member of the unit until he retired, the Guard said.

Guard spokesman Rob Keller said Koenig died near the city of Tikrit, about 100 miles north of Baghdad, while investigating a suspicious object that turned out to be a bomb. Les Koenig, of Aberdeen, S.D., told WDAY-TV that his brother "understood his responsibility as far as why he was over there."

—Associated Press, September 2004

Jeff Kenner, weaving manager for Sioux Manufacturing Corporation (SMC)—a defense contractor that wove Kevlar cloth panels used to mold combat helmets—looked around for something to read while waiting to order parts for his department. As fate would have it, he noticed a condensed copy of the company's contract with Federal Prison Industries, Inc. (also known under the trade name of UNICOR, a wholly owned corporation of the U.S. government under the Department of Justice) to weave Kevlar pattern sets used to mold the shell of the PASGT helmet. This was the combat helmet used in 2004 by the troops in Iraq and Afghanistan. UNI-COR was established by statute and executive order in 1934 to provide federal offenders with opportunities for education and work-related experiences. It makes products for sale exclusively to the federal government, and its products do not compete with private sector companies in the commercial market.

Since Jeff had never seen the contract for the work he was doing, he decided to skim through it. His eyes spotted the military specification for weaving Kevlar cloth. As he read the information, he felt a jolt of shock. His company was shorting the density of weave in the Kevlar used to mold helmets being used to protect soldiers in combat.

For Jeff Kenner this day would mark the start of a major change in his career and his way of life. On September 15, 2004, a week before fellow North Dakotan Lance Koenig gave the ultimate sacrifice to his country, Jeff pulled his well-used 1994 Chevy pickup truck into the employee parking lot at Sioux Manufacturing Corporation. SMC was located in a nondescript single-story building on the Spirit Lake Indian Reservation in Fort Totten, North Dakota. It was a mild, comfortable September morning in this central North Dakota area, where harsh winter weather and hot, humid summers were the norm. The pleasant weather put him in a good mood to begin his daily work routine as he commuted from the nearby town, Devils Lake. Sioux Manufacturing was a good place to work, especially for that part of North Dakota. The benefits and pay were good. He had invested nineteen years of his working career in the company.[1]

Lying in the open plains of the eastern central part of North Dakota is the large body of water known as Devils Lake. Covering about ninety thousand acres stretching over two hundred miles, it's the largest natural lake in the state. The result of an outgrowth of the action of the great Keewatin ice sheet, which pushed down from the north during Pleistocene times, it was the last remnant of the glacial lakes that once dotted the landscape and formed a large part of North Dakota. By 1862 Sioux Nation tribes, collectively called the Dakota, populated the area, along with many other Sioux tribes that settled throughout the great plains of the Midwest from the Rocky Mountains to the Big Woods of Minnesota.

Located on hilly terrain, the Spirit Lake Indian Reservation is dotted with several small towns, the largest of which is Fort Totten, named after a U.S. Army fort built in 1867 to maintain peace in the area between Indian tribes and settlers. At the time the fort housed elements of the Seventh Cavalry that were among those, under the command of Lt. Col. George Custer, killed in action at Little Bighorn in 1876. Today, Fort Totten is the site of Sioux Manufacturing Corporation, the largest employer on the reservation. The tribe also owns and operates the Spirit Lake Casino, which includes a marina, upscale restaurants, a hotel, and a theater. In the summer, tourists flock to the casino but during the cold winter, it is quiet with mostly locals or regional guests at the restaurant and hotel.

After the American Civil War, North Dakota became home to hundreds of thousands of immigrants, mostly from Norway and Germany, who settled the central to eastern part of the state to farm its severe but rich lands. Many of their descendants still live and work in the area. During the late nineteenth century, many small communities sprang up in the lake region; the most prominent town was Devils Lake. This small midwestern town, located about twenty miles north of the Spirit Lake Reservation, is separated from the reservation by a causeway that bridges the lake. It is an isolated town in a cold and windswept part of North Dakota. Its citizens are used to the harsh weather that rakes the area with wintry icy temperatures, at times dipping to a frigid forty degrees below zero. Most of Devils Lake's seven thousand residents are Caucasian; only about 8 percent are Native American.

A natural tension developed between the settlers and the Devils Lake (now known as Spirit Lake) Indian tribe after the February 1867 treaty between the U.S. government and the Sioux—a tension that has continued in varying degrees to present times. The town of Devils Lake and the Spirit Lake Indian Reservation are two different worlds. The causeway that separates these two dissimilar areas represents a cultural divide as well as a physical one.

In 1973 unemployment on the reservation was a staggering 61 percent. The only jobs available were low-paying, unskilled, and seasonal labor, causing most of its residents to exist on federal subsidies.

Through the efforts of local and federal government and the help of then North Dakota senator Milton Young, Brunswick Corporation of Skokie, Illinois, was invited to establish a manufacturing business in Fort Totten. Thus in February 1974, Devils Lake Sioux Manufacturing Corporation was born, jointly owned and operated by Brunswick and the tribe, to produce camouflage or screen systems for the Army.

In 1972 the Army began design of a new, lightweight helmet to replace the heavier, unstable, and uncomfortable M-1 helmet, known as the "steel pot," used by troops for thirty years, from World War II until after Vietnam. In its new design, the Army wanted to use a new and lighter material but at the same time produce a helmet that emphasized ballistic protection and was acceptable to the troops. This effort resulted in the Kevlar helmet known as the Personnel Armor System Ground Troops, or PASGT. The PASGT was constructed of seventeen layers of Kevlar fiber material and laminated with resin. The interior of the helmet contained a webbed-leather suspension system, known as a sling suspension system, that provided a half-inch standoff of the helmet from the skull. Kevlar is a fiber that is approximately five times stronger than steel. The military has used it for bulletproof vests and in high-performance composites for aircraft in addition to the PASGT.

In September 1979 Devils Lake Sioux Manufacturing became one of the first companies awarded a contract by the Army to fabricate 250 test helmets. By 1983 the company was among three contractors to be awarded production contracts to produce the PASGT. At first the company outsourced the weaving of the Kevlar panels and resin coating but later determined it was not cost efficient to do so. By 1984 Sioux Manufacturing began weaving and coating operations to go along with manufacturing the entire helmet shell.

The company experienced a watershed year in 1989—the end of the sixteen-year joint ownership with Brunswick and the end of producing the PASGT helmet. In September 1989 the tribe bought out Brunswick's stock in the company and became the sole owner of the business, adopting the name Sioux Manufacturing Corporation.

By 1989 the Army had nearly three million PASGTs in use and thus eliminated most production contracts through the 1990s, including the one with Sioux Manufacturing. The only contractor to receive production contracts for the PASGT in the 1990s was UNICOR. Because Sioux Manufacturing lost production contracts for the PASGT helmet, the newly owned tribal company had to rely on many smaller contracts, or short orders, from a variety of companies, including Boeing, Lockheed, and FMC, to stay in business. In 1995 it received its first order from UNICOR to provide water-repellent-treated Kevlar fabric for the shell of the PASGT, followed by another contract in 2000 to weave Kevlar cloth into pattern sets used to form the shell of the PASGT helmet.

Having risen in the ranks of employees at Sioux Manufacturing, Jeff Kenner was in charge of the very essential Weaving Department. It was his job to keep the machines running and production moving forward. Jeff had a good mechanical aptitude, a skill that served him well in his job. At the time Jeff may have been the only one at Sioux manufacturing with the skill to keep the big weaving machines running. Unpretentious and genial, he was always willing to help other workers when needed. He was popular with the other employees, mostly Indian tribal members, who usually remained apart from the few white employees.

At six feet, four inches and 255 pounds, Jeff has an amiable, laid-back, friendly nature that belies his imposing stature, earning him the label of gentle giant. Born in 1963 in Devils Lake, he was the eighth of nine children of Darwin and Ruth Kenner. Growing up on the family farm owned by the Kenner family for 125 years, Jeff as a young boy was excited to work in the fields with his father and older brothers, but the excitement faded with age, and he grew to dislike farm work and couldn't wait to leave it behind. After graduating from Devils Lake Central High School, Jeff decided to explore new areas of work. Having discovered that he possessed a mechanical aptitude, he attended the local community college, where he obtained a diesel mechanics certificate and met a nursing student, Shelley, who became his wife. He gained a reputation in the community as a man of principles, a hard worker and family man. As a third-generation Norwegian, Jeff knows no other life beyond this Great Plains community. He has never entertained thoughts of leaving this simple society. He is an uncomplicated but proud man, with a soft voice and given to few words—an example of the Scandinavian Plains culture, in which family, hard work, and loyalty take precedence.

Jeff is also a devout Christian in a town where most of its citizens attend church on Sundays. Jeff doesn't drink. After a bout of drinking in the sixth grade, he "promised God [he] would never drink again." He had to overcome peer pressure, even when tempted at parties, but his faith carried him through. He has a strong, deep sense for what is right or wrong. To Jeff there are few shades of gray.

Jeff started working at Sioux Manufacturing in May 1985. Since Devils Lake is mostly an agricultural community, manufacturing jobs were scarce in 1985, as they are today. With his mechanical skills, he wanted to get into a factory, and Sioux Manufacturing offered the best choice and the highest pay. He started at the bottom of the employee food chain, working as an assembler, surface prepping combat helmets on the graveyard shift, earning an hourly rate of $5.18. His knack for reliability and skill earned him many promotions through the years. The quality of his work did not go unnoticed by management. In recognition of his exemplary performance, Hyllis Dauphinais, the production manager of Sioux Manufacturing, in 2002 presented Jeff with a letter of appreciation for being a committed employee and "supporting the demands of [the company's] daily production schedules." Dauphinais made a point in the letter to let Jeff know that the company gave him more responsibility than others because they knew the job would get done and he would "respond to the company's needs when called upon." By all accounts management thought of Jeff as a model employee.

As he began his daily routine on September 15, 2004, Jeff wasn't thinking about the role that his helmet armor was playing in Iraq, defined by head injuries due to IEDs, the reason Lance Koenig was killed. His immediate concern was the need for more parts for his inventory to keep his machines running to produce the woven Kevlar cloth and stay apace with increasing wartime demand.

To order more parts, Jeff needed to trek over to the material purchasing office. As he normally did, he walked the length of the plant floor to get to the office. When he entered, Tammy Elshaug, the material purchasing manager, was busy peering at her computer. "Could you wait a few minutes?" she asked. "I need to finish looking up some information on my computer."

Never one for wasting time, Jeff decided to read the condensed version of the company's contract with UNICOR, which interested him since the DOD's weaving specification within the contract directly affected his department. He noticed the specified density of the weave was required to be thirty-five by thirty-five yarns per inch minimum (in a crossover pattern). The denser the weave, the more ballistic protection the helmet provided by absorbing the energy from the impact of the projectile. The Kevlar yarn is woven in a lengthwise direction, known as the "warp," and a crosswise direction, known as the "fill." The two weaves together make up the pattern sets installed in a predetermined number of layers between the inner and the outer shell of the helmet. The ballistic performance of the Kevlar helmet depended on the thickness of the Kevlar molded shell.

The number of threads per inch for the fill and the warp is called the "thread count." It is much like the "thread count" of woven linens and bedsheets, indicating the number of horizontal and vertical threads in one square inch of fabric. As he read the specification, he could barely comprehend the significance of what he was reading, but then it hit him like a thunderbolt. He had always thought that the density requirement of Kevlar threads was thirty-four by thirty-four yarns per inch or less. It was what they had routinely been weaving for years, since the mid-1990s at least. He recalled when in the early nineties, his boss, the weaving manager at the time, had directed him to change out the reeds on all six weaving looms so they could weave at thirty-four by thirty-four. Jeff had asked him at the time why he wanted to change out all the reeds. It was a lot of work and seemed like a waste of time. "Because they are going to save a pile of money," was the reply.

Along with weaving Kevlar cloth came the tedious task of counting fibers. Counting threads was a difficult process under normal circumstances, requiring a great deal of focus by the worker. The cloth moves through the machine at an angle, the machine shakes, the noise, the pressure to get the count right all make it very difficult to concentrate on the count.

Jeff knew enough about military specifications that they were in the contract for a reason. Military contracts for products ranging from corn flakes to supersonic jet fighters come with specifications or requirements that the product must meet. The need for specifications has been an obsession of the military since the American Civil War. Mil specs, as they are called, have to be developed for every product imaginable, even for tools that can easily be purchased at a local hardware store. However, mil specs are crucial for important war fighting equipment, such as troop protective equipment, and are developed after much research and testing. Often the delivery of the equipment is accompanied by a certification, signed by the manufacturer under penalty of federal prosecution, that the product conforms to all specifications required by the contract.

In this case the weave of the cloth had to be dense enough to ensure that the helmets protected the soldiers' heads from small arms and shrapnel, including shrapnel from an IED. If the military thought that the thirty-five-by-thirty-five-yarns-per-inch weave was the minimum needed to protect the soldiers, Jeff felt the company needed to comply to safeguard the soldiers. His mind turned to the final destination of the helmet: the ongoing wars in Iraq and Afghanistan. He wondered if what his company did to save money could be putting troops in peril. Would the shorting of the weave allow projectiles to fully penetrate a helmet? He recalled seeing the results of helmet tests at Sioux Manufacturing and the large bubble that formed on the inside of the helmet after projectile impact. Could that be because shorting the weave weakened the protective value of the Kevlar and resulted in more serious impact to a soldier's head? He didn't know for sure, but he believed that it had to make the helmet weaker.

Whether it had a combat impact or not, Jeff felt it was unethical to be shorting the weave by 2 to 5 percent per helmet, which amounted to 713 feet of missing Kevlar yarn for each helmet. He believed Sioux Manufacturing had a contractual obligation to the military and the troops to do it right. Despite Jeff's laid-back and quiet demeanor, he is a stickler for integrity, for honestly doing a job. He decided to make sure the problem was corrected as soon as possible and that meant reporting the violation to Carl McKay, the company CEO, and voicing his concern. Jeff had always been a "team player," loyal to the company at every turn. His decision to complain to McKay about this contractual violation was an easy one. He had direct access to him. If there was a problem, Jeff would always go to him—he never thought twice about it. McKay was used to Jeff coming to him with internal problems. However, this time, unbeknown to Jeff, it would be different. This type of complaint would label Jeff a dreaded "whistleblower."

Carl McKay was born in 1948 at Fort Totten, attended the University of North Dakota, received a bachelor's degree in social work in 1971 and later a law degree in 1984. He became tribal chairman at the age of twenty-six and served in that capacity for seven years before becoming CEO of Sioux Manufacturing in the early 1980s.

Jeff thought that going to McKay with any issue, even with the disclosure of the practice of shorting the weave, would not be difficult. He expected McKay to embrace his discovery, understand the problem, and take steps to rectify it. It is not unusual for employees who have discovered internal company practices that violate the terms of a government contract to feel they are doing the right thing by reporting such practices to their boss or even top-level management. They often believe they will be praised for doing so. Normally, they don't consider themselves whistleblowers but rather think they are just being good employees.

However, violating the terms of a government contract can have the effect of a contractor or subcontractor losing its contract to the detriment of the company's future viability. Thus upper-level management may feel threatened by the disclosure and, instead of praising an employee for bringing the problem to its attention, take action to discredit, demote, or otherwise harass the employee with the objective of terminating the person's employment. This was the scenario in which Jeff unwittingly found himself by reporting the weave shorting practice to the CEO.

McKay was not the typical CEO. He often walked around the plant telling jokes to the workers and was generally liked. Even with McKay's easy-going demeanor, Jeff was leery of his style. At one time he had respected McKay, but the more he dealt with him the more Jeff realized the CEO would just tell someone what he or she wanted to hear and wouldn't really act on it. McKay's heritage is Native American, but unlike most of the Native American workers at Sioux Manufacturing, who lived on the reservation, McKay lived in New Rockford, a town forty miles south. He was one of only a few Native Americans at the time to attend Central High School in Devils Lake.

Carl McKay and the other top executives of Sioux Manufacturing were sequestered in a complex of offices in the plant ironically known as "the Pentagon." The complex is kept locked and is not normally accessible to employees without an appointment or knowledge of the door-lock code. Jeff knew the lock code and could get into the Pentagon whenever he wanted. As he normally did, he walked into McKay's office, anticipating a resolution to his concern, not suspecting a pushback.

"Got a minute?" asked Jeff, unconcerned. "Ya, sure," McKay replied. Jeff sat down and got right to the point: "Why were we told to weave the cloth at thirty-four by thirty-four?" Usually McKay was mellow, almost matter-of-fact in his response. But Jeff saw him stiffen, his demeanor suddenly changing. This was not normal, Jeff thought. He wondered if he had told McKay something he shouldn't have. "That's the way we are going to weave it. Don't worry about it," McKay sternly replied. Jeff objected, "I don't think it's right." McKay quickly reprimanded him: "Well, I'll be the judge of what's right."

McKay's reaction was a bolt out of the blue for Jeff. He was taken aback, bothered by McKay's rebuke of his attempt to resolve what Jeff felt was a serious problem. It was a peculiar change from McKay's usual manner of placating an employee with a complaint even if he didn't act on it. As air in the office thickened with tension, Jeff scoured his mind, trying to figure out why McKay was reacting in this way. His discomfort grew, and the uneasy feeling that he had overstepped some invisible corporate line with McKay washed over him. Jeff thought he was doing the right thing, conveying news of a problem that needed to be solved. However, McKay offered a more modest recollection when interviewed for this book nearly seven years later, claiming that Jeff only had an "engineering issue," and he (McKay) couldn't recall the details.

Leaving McKay's office, with his head reeling and his heart beating rapidly, Jeff quickly walked back to Tammy Elshaug's office. He needed to unload on someone about what he had just gone through. He trusted Tammy and didn't hesitate to confide in her what he had discovered regarding the weaving problem and how disappointing his attempt to resolve it with McKay had been. "He just blew me off, that it was none of my business," Jeff told her. He showed Tammy the language of the contract and explained his belief that the company was violating the contract by shorting the military of the required amount of Kevlar. She was as shocked as he was and had the same gut fear that they could be fielding unsafe helmets to the combat troops. They agreed something needed to be done about the problem because of the potentially fatal consequences to the troops.

Tammy had always believed that the thread count was thirty-four by thirty-four yarns per inch. It was something that was talked about around the plant. Before Jeff told her about the specification in the contract, she had been uncertain what thirty-four by thirty-four actually meant in terms of the requirements but had always thought it was the standard. She periodically calculated the amount of Kevlar that should have been used in producing the number of pattern sets delivered to UNICOR. Since Kevlar was expensive, it was important for her to know exactly how much Kevlar to reorder from DuPont, the Kevlar thread supplier. What had mystified her was why she would always end up with a surplus.

When Tammy submitted inventory reports to management showing surplus Kevlar, the managers would always make her and her employees feel as if it was their fault the numbers were not adding up right. After Jeff's revelation of shorting the weave, she realized the reason for the excessive Kevlar. She reconfirmed her new understanding by conducting a physical inventory with Jeff shortly after his discovery. Again, they came up with a surplus of Kevlar along with a shortage of resin. The resin shortage indicated to her that they were using extra resin to make up for the lighter weight of the pattern sets resulting from weaving less Kevlar then required. More resin would ensure that the product met the weight specifications required by the military, thus compensating for the shortage of Kevlar, but it would not make up for the lack of protection.

Born Tamra Schlieve in Devils Lake on August 24, 1958, Tammy was the fourth of six children in a middle-class family. Her father was a military man who served in the North Dakota National Guard as a warrant officer specializing in mechanics at Camp Grafton, located outside Devils Lake. Her mother taught elementary school in the small town of Tokio on the Spirit Lake Reservation.

Tammy's great-great-grandparents had immigrated to the Devils Lake area from Norway and Germany. Like Jeff Tammy attended Central High School in Devils Lake. While in the ninth grade, she started working as a waitress and other odd jobs such as a sales clerk at the local grocery store. After graduating from high school, she didn't want to marry and be a young housewife like so many of her friends, so she enrolled at Lake Region Community College, where she majored in liberal arts. She subsequently attended Aakers Business College in Grand Forks, North Dakota, where she received a clerical certificate in 1978. After returning to Devils Lake, she took a job at Lake Region College as a secretary for a few months.

Her college secretarial job didn't pay much and was uninspiring to boot. In April 1979 Tammy learned of an opening at Sioux Manufacturing (then known as Devils Lake Sioux Manufacturing Corporation) and applied for a production job, but instead she was offered a secretarial position in Quality Assurance. Disappointed but seeing some possibilities for advancement in position and pay, she took the job. She remained in the quality assurance position for about two years, producing progress and scrap reports while learning the inspection process. Tammy noted that most people in management at the time were Brunswick employees who had transferred to Fort Totten. Others in management positions were people who were not from Brunswick but had college degrees in fields such as engineering. "Brunswick was more demanding in wanting to do things right. More discipline, structure, and organization. It was run like a real production house," Tammy recalled.[2]

Tammy felt that Brunswick ran the plant in a professional manner, implementing many policies and procedures. "They took the policies from their parent corporation and applied them to Sioux Manufacturing until the tribe took over." When the tribe took over, "everything was laid back." After Brunswick left, the tribe hired professionals but tried to keep tribal personnel in key positions. "We kept doing what we were doing, but there was not the pressure from the corporation," said Tammy.

Eventually, Tammy was promoted to executive secretary to the general manager, but she remained in this position only a couple of months. Like Jeff she moved up in the company because of her diligence on the job and achieved the position of materials purchasing manager in 1997. She was the first woman in the company to achieve that position and was given the responsibility of buying material for the entire facility.

In her new job, Tammy handled all material requests and kept tabs on material inventory. She was responsible for making sure the production departments were well stocked with the materials needed to maintain tight production schedules—a critically important position. Her supervisors noted in her reviews "knows job well," "excellent at meeting challenges and problems," "very good problem solver," and "excellent attitude and team player." Typical were comments from one of her supervisors: "She has to be considered one of SMC's most valuable employees.... She not only performs her duties very well from the internal materials functions but is instrumental in much of the traffic activities inherent to SMC's operations." By September 2004, with twenty-five years of employment with the company, Tammy was considered a respected asset to Sioux Manufacturing.

In 1995 Tammy married Larry Elshaug. He worked for a construction company that specializes in concrete work—"a lot of paving," explained Tammy—and spent much of his time on projects at Minot Air Force Base pouring runways. They had two boys, who were very active in ice hockey. During the winter months, Larry erected a small hockey rink next to their house for the boys to play and practice on. Tammy was a true hockey mom who shepherded the boys to hockey practices and games. The Elshaugs lived in a modest home not far from the downtown center. As with the Kenner's, Devils Lake had always been their home. Like their immigrant ancestors, their lives were taken up with work, family, and tradition.

Both Jeff and Tammy, according to one manager, "had good reputations at the plant."[3] All that hard work, goodwill, and loyalty, how-ever, would not protect them after they questioned the gap in the weaving program because it threatened the company, its contracts with the military, and the profits for the tribe, which relied heavily on the company for its sustenance. In their naive view of the world, Jeff and Tammy thought that telling McKay was the logical and right thing to do for the company and the troops. Yet they didn't under-stand the complicated world of military procurement or the socio-logical tensions in their own small town that would lead them down a difficult path of trying to tell the truth and surviving unscathed. With this disclosure their problems had just started.

CHAPTER 4

Blowback

One of the biggest mistakes whistleblowers make is assuming that management will recognize and praise their disclosure and quickly resolve the issue. In reality that rarely happens. More likely management will feel threatened by the disclosure because of the potential loss of contracts and revenue, resulting in devastating consequences for the company. Management's reaction almost always takes the form of "shoot the messenger."

Management does not want to admit that its reaction to the employee whistleblower is due to the disclosure. Instead, it takes a different tact with the objective of terminating the employee. To achieve that objective, performance problems, along with efforts to discredit and ridicule the employee, are usually created, resulting in demotions, transfers, harassment, poor evaluation reports, and so on. This campaign usually results in forcing the whistleblower into termination, either voluntarily or involuntary. Management will "keep book" on the whistleblower, documenting everything it sees and hears, including "company gossip" whether true or not. Jeff Kenner and Tammy Elshaug would find out how management views whistleblowers despite their value to and longevity with the company.

Starting in 2002 and continuing through 2006, orders for woven Kevlar pattern sets increased sharply in response to war in Iraq and Afghanistan. As weaving manager Jeff was responsible for keeping the Weaving Department running smoothly, continuing to move company production forward. Weaving was a difficult department to manage. Its six machines were sophisticated, and if any of them broke down or just needed a minor repair, production was curtailed. When this happened, and it happened frequently, Jeff diagnosed the problem and fixed it.

In 2003, when Jeff took over as weaving manager, the department was barely running; machines routinely broke down, which almost halted production and seriously impacted revenues. Thus Jeff was put in an unenviable management position, entrusted with responsibility for resolving its problems and getting the department running smoothly again. This task took considerable effort and a lot of overtime to accomplish, but Jeff managed to do it, increasing production just when the company needed it most.

Nobody around the plant would actually say that Jeff was doing a good job. It wasn't the culture of the plant, and Jeff never expected it. Everyone just knew that he did a good job. Jeff managed twenty-six employees on three shifts. It was a tough period as he would frequently be called during the evening or in the middle of the night with mechanical issues in the department, but he would come into the plant and do whatever it took to resolve the problem. During 2003–4 Jeff made the third highest wage in the plant because of the hours he had to put in. The workers in the Weaving Department liked Jeff a lot, especially since he always fought for them to get raises. Management expected Jeff to also teach the newer workers how to operate the machines. He was always there to mentor them and was very patient even though at times he wanted to pull his hair out.

It didn't take long for Jeff to realize what he had stirred up with McKay. On September 20, 2004, five days after reporting the weave shorting problem to the CEO, Hyllis Dauphinais, who had recently been promoted to operations manager, called Jeff into his office and announced that he was no longer the weaving manager. Jeff was being demoted with a transfer to the Maintenance Department. Jeff was shocked and confused. He asked Dauphinais why he was being demoted, and the explanation was to give another employee with more seniority a chance to be weaving supervisor. Jeff, highly skeptical of that explanation, replied he did not want to work in maintenance because the people there were "nothing but backstabbers." He appealed to Dauphinais, but Dauphinais held fast to his decision, condescendingly telling Jeff it was "for the good of the company."[1]

The next day, while still greatly agitated about the demotion, Jeff appealed to McKay. After a long discussion, McKay rescinded the demotion, agreeing to allow Jeff to continue managing the Weaving Department. Feeling that the matter was resolved and he could resume his position as weaving manager, Jeff composed an email and sent it to McKay, thanking him for addressing his concerns and agreeing to keep him in weaving, "giving [him] the privilege to run it."[2]

About an hour after receiving the email, McKay replied that he had changed his mind and would allow the demotion. "He [Dauphinais] also needs to correct an inequity relating to seniority in the Weaving Department," McKay explained after discussing the matter with Dauphinais. "I failed to follow our seniority policy and overlooked the fact that another employee has more seniority than you and that employee was not given an opportunity to benefit from our established policy. My failure to recognize this fact was not fair to another employee."[3]

Jeff was distraught but resolute. After all his many loyal years with the company, he was not going to take a demotion lying down, so he was not going to work in the Maintenance Department. Soon Jeff learned that the weaving manager position did not go to a more senior employee because that employee didn't want the promotion. Instead, it was given to another employee with less seniority than Jeff. Now certain that his demotion was retaliation for bringing the weave-shorting issue to McKay's attention, Jeff met with both McKay and Dauphinais. He was armed with McKay's email in hand that falsely stated that his demotion was due to the company seniority policy. He showed them the email and asked to be sent to the Materials Department instead of maintenance. They quickly agreed. Jeff was officially transferred to the Materials Department but with a hefty reduction in pay.

The official explanation by management would later contend that Jeff was not "demoted" but had been only "temporarily" transferred to the Weaving Department the previous year and was simply being moved back to Materials at his prior hourly pay rate. It was part of the choices made by Sioux Manufacturing in hiring "lead" and "supervisory" weavers and "maintenance weave technicians" and was unrelated to Jeff's claims made. However, the human resources (HR) manager had a different view. He believed it was the result of a dispute Jeff got into with top management. In a different take on the issue, the HR manager recalled Dauphinais telling him that Jeff was being demoted for being "uncooperative" but did not elaborate.[4]

The next day Jeff transferred out of the Weaving Department and began working as a materials clerk under Tammy Elshaug. Many plant employees expressed confusion about why this change had taken place. Jeff was well known for his expertise and mechanical abilities, a skill set that no one else in the plant had to keep the machines running in the Weaving Department. It was common knowledge throughout the plant that Jeff was the person who had kept that area running efficiently. His demotion didn't make any sense to the people on the plant floor.

As predicted by many employees at the plant, the Weaving Department went downhill not long after Jeff was removed. The machines broke down and stayed down. The department could barely maintain three recently purchased weaving machines. The company had to purchase an enormous amount of expensive machine parts to repair breakdowns caused by human error. Overtime became common to compensate for slow production due to the unreliable and disabled machines. Jeff was no longer there to keep those machines running, and no one else seemed able to do so. The HR manager tried to talk McKay into putting Jeff back into weaving, but he refused.[5]

Indeed management was not through punishing the whistleblowers. Three weeks later, on October 19, 2004, Dauphinais called Tammy into his office. To her surprise he told her there would be a reorganization in her department. He was going to make Julie Sailor, who was a subordinate of Tammy's, responsible for the Purchasing Department and make Skip Longie Jr., Dauphinais's brother-in-law, the materials manager responsible for both the Purchasing and the Materials Departments. That meant Tammy was being removed from her managerial position and demoted. Dauphinais explained he wanted to use her expertise somewhere else. Tammy sat there listening to him, steaming inside, but she remained silent. Dauphinais told her that he was going to make her a warehouse supervisor. She struggled to grasp what was really going on. It didn't make any sense to her. The longer Tammy sat there without saying anything, the more it seemed to bother Dauphinais. Finally, he said, "Say something, tell me what you're thinking."[6]

Finally, Tammy spoke up. She asked Dauphinais why Longie was taking over her job. His explanation was that Longie brought fifteen years of warehousing experience. Tammy then asked him, "What about my twenty-five years of experience. Doesn't that mean anything to you?" She asked him if he thought she wasn't doing her job. He assured her it was not that but because it was "for the good of the company." That was the tipping point for Tammy. Red-hot angry she blurted out to him, "If I was you, I'd think twice about what you're trying to do to me because I know you're not in compliance with the weaving of the Kevlar cloth for the UNICOR contract." Dauphinais became irate and mumbled something unintelligible, but Tammy stalked out of his office and headed straight to McKay's office. Tammy repeated to McKay what Dauphinais had said about the "reorganization." In order to appease her, McKay said he was not aware of the decision and told Tammy not to worry about it.[7]

For both Jeff and Tammy, the tenor of company management's reaction and subsequent retaliation would permanently alter their years of faithful employment. Their loyalty to the company had been shattered. Their loyalty was now to the troops who had to wear the helmets.

The reorganization never took place. Tammy didn't know why. It could have been Dauphinais backing down, McKay nixing it, or something else. She never heard any more about it and continued working as the materials manager with Jeff as one of her subordinates. However, their revelations about the shorting of the weave did not end with McKay and Dauphinais. At subsequent meetings within the Quality Assurance Department, both Jeff and Tammy voiced their objections to what they felt was systematic shorting of the Kevlar bound for combat helmets. It wasn't long before their objections resulted in more hostility from management, mostly directed at Tammy in the form of continued verbal harassment.

With continuing hostility from management, Jeff decided to retreat from trouble. He didn't complain further about the weaving compliance issue, not even to Dauphinais, lest it bring him more trouble and possibly cost him his job. Jobs were scarce around Devils Lake. Working now at the level of a clerk in the Materials Department was a real step down from Jeff's management responsibilities in the Weaving Department. His office was a closet in an area he had cleaned out. Tammy was now his boss, and he took the opportunity to learn as much as he could about how the department worked and how inventory was maintained.

Suddenly out of the blue, a rumor began behind their backs that Jeff and Tammy were having an affair. This rumor was floated around the plant shortly after both of them made their complaints to management about the weaving issue. A coworker friend of Jeff's took him aside and said that one high-level manager was telling certain workers that Jeff "was a real piece of shit" and "was having an affair with his boss, hidden behind closed doors all the time."[8]

Jeff was alarmed by this news. The gossip wasn't true. However, it was common for workers at the plant to spread rumors about workers having affairs, and these rumors had a tendency to take on a life of their own. It hit home that he was now the target of such rumors. Embarrassed because she was his boss and friend, Jeff decided not to say anything to Tammy about it since she was unaware of the rumors at that point. Another worker teased him about the affair, but it wasn't funny to Jeff. Staring directly into the worker's eyes, he asked him, "Are you my friend?" "Yeah," replied the worker. "Do you want to continue to be my friend?" "Yeah," was the reply. "Then cut the bullshit about Tammy and me," Jeff told him. "It's bullshit and it's not helping anything by you spreading it further." The worker grasped the seriousness in Jeff's words coupled with his imposing physical presence and took the opportunity to walk away.

Despite Jeff's efforts to stop the rumor by denying it to anyone who brought it up, it spread around the plant like wildfire. Additionally, Dauphinais started treating him "like dirt by belittling [him] and always [getting] on[his] case." Dauphinais was "using a more hostile tone toward [him] than before." This abuse was not limited to Jeff. Dauphinais started making demeaning comments to Tammy as well, especially at meetings they attended together. It wasn't long before she heard about the rumored affair.

CHAPTER 5

Whistleblower's Nightmare

On November 10, 2004, the rumored affair between Jeff and Tammy took a new turn. Tammy's husband, Larry Elshaug, was driving his company pickup truck from Devils Lake to Minot, North Dakota, where he would be managing paving projects on an Air Force base. He was about three-quarters of the way to his destination. Suddenly, his cell phone rang, jarring him out of his thoughts of the day's business. A woman's voice, sounding somewhat familiar, was on the other end of the call. The woman did not identify herself but said she worked with his wife and there was something he should know; Tammy and Jeff Kenner were having an affair. From the voice Larry guessed it was Julie Sailor—one of Tammy's subordinates—because Tammy had mentioned to him more than once that she had a reputation for gossip. He again asked her for her name, but she would only say she was just somebody concerned and that he should know there might be something going on. Larry persisted in asking her for her name. Relenting, she finally said it was Julie. "Is it Sailor?" he asked. "Yes," she replied. The phone number she had called from indicated to Larry that it came from Sioux Manufacturing. He told her he would look into the matter and ended the call. He made a note about the call in the log he maintained of his daily activities.[1]

When Tammy heard from Larry about the call, she was shocked by Sailor's deed. Sailor worked for her and they appeared to get along, but she knew Sailor was obsessed with gossip. Sailor had always acted that way, but now she had called Tammy's husband with boldface lies. There was no doubt in Tammy's mind that it was Sailor who had called him, and Tammy believed management had put her up to it. The next day, visibly upset, she confronted Sailor at work. Crying during the confrontation, Sailor denied making the call, but Tammy sensed that she was not telling the truth. Tammy sought the advice of the HR manager, who suggested she file a formal complaint against Sailor. Tammy did, and the HR manager called Sailor into his office and admonished her, but she continued to deny that she had made the call. McKay and Dauphinais supported Sailor's side of the story and would not do anything about it. She eventually submitted her own complaint against Tammy for slander and defamation of character, but nothing came of it.[2]

The affair rumor soon began to spread throughout the small, tight community of Devils Lake. As a result, a strain on both Jeff's and Tammy's marriage erupted. Along with a pattern of hostility by Dauphinais and other employees at Sioux Manufacturing, the rumor began to take a toll on Tammy. She was startled after a tree-planting dispute when a maintenance worker threatened to cut her head off. In another incident Dauphinais screamed at her over a minor delivery issue, causing Tammy to flee the office.

Tammy had not experienced this type of behavior before during her twenty-five years of work for the company. She now dreaded going to work, worrying that management was trying to set her up to fire her. By August 2005 she was depressed and felt she could not work for the company any longer under this tension and abusive treatment. She strongly believed Sioux Manufacturing was defrauding the government, cheating the taxpayer, and putting soldiers in danger. Not sure how to deal with the dilemma of needing the job and putting up with the harassment and the toxic rumors, she decided to keep her head down, walk on eggshells, and avoid confrontation.[3]

By early October 2005, more than a year after Jeff had revealed to McKay that the company was not in compliance with the weaving specification, Dauphinais continued to verbally harass Tammy, while the affair rumor still circulated around the plant. By then knowledge of the weave-shorting issue had spread among many workers. "The cat was out of the bag," Jeff thought. "It was all over the plant." At about this time, Rhea Crane, the quality assurance (QA) manager, became aware that the weave was not supposed to be thirty-four by thirty-four yarns per inch. It hadn't occurred to her that the specification called for a minimum of thirty-five by thirty-five yarns per inch until a company QA inspector pointed it out to her, and she verified it by reading the specifications after obtaining a copy from the contracts manager.

Looking into the issue, Crane found that Kevlar woven panels were being produced with yarns per inch as low as thirty-two by thirty-four and thirty-three by thirty-four. As a result she put a hold on many rolls of the woven cloth, not wanting to put her name on it if shipped. She met with Jeff and Tammy and gingerly acknowledged the problem. "I didn't know the spec called out for thirty-five by thirty-five minimum," she admitted, thinking the spec was thirty-five by thirty-five, minus one. Despite this revelation Jeff later determined that the Weaving Department continued to produce pattern sets that had been woven at thirty-four by thirty-four because it would take a change of certain parts, called "reeds," in each weaving machine to weave at a higher density. He also did not see any action on the part of management to change to the proper weave specification, especially because it would be a long and slow process.[4]

On October 24, 2005, a little more than a year after Jeff had discovered the shorting of the Kevlar weave, he noted that the shorting still had not been corrected. He confronted Dauphinais by asking him if he was "trying to save the plant even more money by running even less pick counts." "No, someone in weaving is making mistakes," was his reply. "That's highly unlikely," Jeff responded. "Somebody's got to hear about this." Jeff stalked away. He was fuming and had hit his limit.[5]

The next day Tammy sent Jeff into Devils Lake on a routine task to purchase supplies they needed for the department. It was the middle of the afternoon, and Jeff took the company truck and a bag of cash, about $200–$300, to pay for the supplies. It had been a rotten day for Tammy. She had met with Dauphinais, and it was the "same old stuff"—with him demeaning, accusing her, pointing his finger at her. Finding herself irate and unable to concentrate on work, Tammy decided to go home early. She couldn't take any more abuse.[6]

Jeff picked up the supplies and drove back to the plant. As he drove along Highway 57 on the reservation, less than a mile from work, he spotted Tammy driving in the opposite direction. Likewise, she saw Jeff in the company truck. She was angry and stressed-out from the tensions at work. She quickly decided to signal him to stop. They needed to talk privately, away from the company and from the prying eyes and ears of coworkers. He pulled over to the side of the road where Tammy had stopped, at the intersection of BIA Road 24 and Highway 57. Jeff immediately could tell something was terribly wrong. She started crying, and he decided to find a private place to talk. They were on a heavily traveled highway, and he didn't want her to be seen crying by passing motorists, especially company workers who might be driving by. Tammy followed him down Fortieth Street NE through a residential area where kids were playing. Not an ideal place to stop, they drove further until they located an empty grassy lot, less than a mile from Highway 57.

Pulling off the road, they backed into an area near the trees, their cars side by side, still clearly visible from the road. An old gate still stood on the lot to mark the existence of a residence, long gone. Both remained in their vehicles and talked through their respective windows. Jeff asked her what was wrong. Tammy again broke into tears. "I just can't take the harassment from Hyllis [Dauphinais] any more. I'm so tired of him treating me like this," she told him. Several cars drove by as they were parked on the roadside. They lived in a small community where "everybody knew everybody," and she did not want people to see her crying. Tammy suggested that they both move their vehicles farther away from the road.

Jeff was at a loss of what to do. He was beginning to think that exposing the weaving problem might not have been worth it. The harassment and abuse Tammy was experiencing, the ugly rumors, were all adding up and making it difficult to manage at work. "Tammy, I'm really sorry I got you involved in all the problems at work," Jeff said. "Maybe we should just blow the whistle and report everything to the Feds. We'll probably lose our jobs, but this has gone on too long."

After a few minutes as they continued to talk, a reservation fire department pickup truck with two firemen appeared on the road and pulled into the area where they were parked. The truck didn't stop, but slowly moved by them, then turned around and slowly made its way back to the road, where it stopped. Knowing they were spotted and concerned the firemen might think it was odd they were there, Jeff decided to drive to where they had stopped to explain their presence to them.

Jeff inexplicably told the firemen he was on his lunch break. He asked them if they were going to burn in the area because they routinely burned grass for fire control. The firemen acknowledged they were. "Well, I'm glad you saw us," Jeff told them. Tammy remained in her car; she had not moved. Because she had been crying, she ducked down to shield her face from the firemen to avoid embarrassment. Then the firemen drove off, followed by Jeff and then Tammy. She continued toward home, while Jeff drove back to the plant. Neither thought much about the encounter—it was merely to talk out their concerns about the hostile work environment that had been created around them, to understand it and how to deal with it. However, management chose a different interpretation of their meeting in this private spot.

Unbeknown to Jeff and Tammy, one of the firemen in the truck apparently notified Sioux Manufacturing management soon after the incident that they had located a company truck along with a private vehicle in the remote grassy area along with the two individuals who were in the vehicles. What would normally have been a minor incident resolved with a verbal notification, the fire chief Daniel Herman, who was one of the firemen in the truck, was asked by Sioux Manufacturing management to submit his version of the incident in writing.

The next day, November 1, Herman provided Carl McKay with a typed statement acknowledging that the firemen had spotted both a car and a pickup truck parked at the location of a former residence. The statement said, "[We] didn't see anything other than the vehicles parked at this former residence, and it appeared that nobody was around. When we left, the pick-up drove toward us and stopped us on the road. The driver of the pick-up stopped us and said 'Ok you caught us, but it's all good because I am on lunch break.' We asked him who's [sic] truck was he driving, but his only response was 'it was all good.' After a few minutes the other car came out of the residence and that was that."[7]

Although the firemen had not seen anything unusual going on in the vehicles or even seen anyone at all, McKay, for whatever reason, interpreted a clearly written statement in a different way. "On a weekday afternoon at approximately 3 p.m. in early October 2005, the two were discovered in a remote woody area of the Reservation by two members of the Fort Totten Rural Fire Department," he wrote. "The fire department members observed two vehicles (one was an extended cab pick-up truck with an SMC decal on it, marked for official use only; the other was a sedan)." He continued with a description of events clearly different from those described by Herman: "Mr. Kenner was observed leaving the back seat of this extended cab pickup, and entering the front seat of the pickup on the driver's side." Implying the meeting between the two was something that it was not, he continued, "The firemen reported this information to the fire department, and the news of this 'tryst' rapidly became common knowledge throughout the Spirit Lake Reservation. The use of SMC vehicle and the fact that Ms. Elshaug was Mr. Kenner's direct supervisor at the time was detrimental to morale at the SMC plant."[8]

Following the Herman statement submitted to McKay, the HR manager was called into Carl McKay's office on the same day. McKay told him that he was terminating Jeff and Tammy because they were involved in immoral conduct on company time, and with all the rumors, it was not good for the plant. The HR manager tried to talk him out of it without success, and he returned to his office very troubled by this sudden turn of events to terminate two valuable employees. He met with Jeff and explained to him that new and disturbing rumors were circulating about Jeff and Tammy. Apparently one of the firemen in the truck had called someone at the plant and said that they had caught Jeff and Tammy in a compromising sexual situation. Jeff reacted immediately, "It was bullshit." The HR manager readily agreed. Then he nervously broke the news: "Gosh, I'm sorry. Carl's firing you." "No way," Jeff responded in disbelief. The HR manager asked him what had happened out there. "We weren't doing nothing out there," Jeff insisted. "But Jeff, you are being terminated for some reason. I've talked to him [McKay] and there is nothing I can do. I tried talking to him," The HR manager explained.[9] Jeff was dumbfounded. After twenty years this was how it was going to end? His immediate impulse was to confront McKay.

Jeff quickly walked to McKay's office, where he found him sitting at his desk. Still in shock over the news, he came right to the point: "Why are you firing me?" McKay replied in a nonchalant manner that since it was an "at-will" state, the company did not need his services anymore and could thus let him go. He gave Jeff no other reason. "It's wrong," Jeff responded, not comprehending the reason McKay was giving him. With his mind in a jumble, all Jeff could think of was to ask McKay if the termination could be delayed. McKay replied, "No, it's better to just go through the motions. Pack up your stuff, and I'll let you know." McKay then mentioned that the rumor of the affair between Jeff and Tammy had "spread everywhere." "You know that's bullshit," Jeff retorted as he walked out of the office in a daze.[10]

The HR manager then met with Tammy in his office and gave her the bad news that she was being terminated. Like Jeff she went directly to McKay's office and demanded to know the exact reason why. She was livid and confused. McKay told her she was not being terminated; she was just "being let go." He then gave her the same explanation that he had given Jeff: North Dakota was an "at-will" state, and the company did not require her services anymore. McKay was dismissive of her and the termination and downplayed its importance, that it was just a routine matter. Both Jeff and Tammy picked up their possessions and left the plant, still unsure of why they had been terminated.[11]

Sioux Manufacturing never revealed the source of the information for the company's version of events that was not contained in Herman's eyewitness statement. The HR manager was just as perplexed about the termination as Jeff and Tammy were. He thought management was trying to get rid of them but did not know the real reason for this action against two valuable, long-term employees. It was only after they had left their jobs that management directed the HR manager to use a violation of the employee handbook, rules number 40, "No employee shall engage in immoral conduct or indecency," and number 45, "No employee shall engage in conduct off the Plant premises or during non-working hours which affects the employee's relationship to his/her job, his/her fellow employees, his/her Supervisors, or the Corporation's products, property, reputation or good-will in the community," as the reason for termination.[12]

However, according to the HR manager, management presented no evidence to him justifying termination for violation of the employee handbook rules. It was too vague, and the Human Resources Department was not allowed to be vague about anything in a personnel file. The HR manager asked Dauphinais what kind of evidence the company had because he was afraid it was going to blowback on the company if Tammy and Jeff retained attorneys. His only reply was, "They were caught by the Fire Department, and there is a statement." The HR manager never saw the statement. He warned Dauphinais that they could not terminate Jeff and Tammy in that way; they had to go through proper procedures. Dauphinais, however, offered no further clarification, and the HR manager did not want to cause trouble for himself by questioning him further. He felt that other employees were now scared to speak out because they did not want to put their jobs at risk, especially seeing how quickly management had been willing to terminate two longtime employees.[13]

When Dauphinais was later interviewed for this book, he claimed that he was not involved in Jeff's and Tammy's termination but believed there was "supposedly some impropriety relationship, or something, that was going on."[14] McKay, during his interview for the book, insisted that an improper relationship existed between the two, that Jeff was "playing footsie with a married woman; the scuttlebutt [of the affair] was within the company." Although McKay acknowledged there was no proof of an affair, he held fast to the reason for the termination, that they were "terminated because the affair was a violation of company rules and you can't have that conduct going on." Despite information to the contrary in Fire Chief Herman's statement, McKay maintained that they were caught in the backseat of an extended-cab truck with Jeff getting out of the truck pulling up his pants. In trying to reconcile the disparity between the fireman's report and his own interpretation of the event, McKay claimed that the "implication was there."[15]

After the termination Jeff drove away from Sioux Manufacturing in a state of shock. He was distressed on the drive home but then just became numb with disbelief. Many thoughts were going through his head—one being to exact some sort of revenge, but he knew he would never act on those thoughts because it was not in his nature. Tammy was also angry at the termination, but her feeling was tempered somewhat by a sense of relief that she no longer had to deal with the compounding stress of working at Sioux Manufacturing. She still didn't know why she had been terminated; that fact would not be disclosed to her until it showed up in her termination papers a few weeks later. She was mortified that the company had put in writing that she had acted immorally, but she felt helpless to change it. Despite the termination Jeff felt he needed to somehow blow the whistle on Sioux Manufacturing for its shorting the weave and possibly putting troops in danger with a helmet that might not adequately protect them. That became his new mission.[16]

CHAPTER 6

The Tinkerer and His Unique Foam

I t started with a headache. It was 1989, and Mike Dennis was flying his private plane from Scappoose, Oregon—a picturesque
town on the Columbia River, about thirty miles north of Portland—to Pennsylvania, with his wife, Jude, to see her mother. On
the long flight, the headset she wore sitting alongside her pilot husband gave her severe headaches. It felt as if someone was drilling a
screw into her head. She tried to put rags underneath her headset,
but it did little good. The headset hurt because it was a one-pound
weight sitting on a fairly small area. Mike thought about it, that the
top of her head must be like a baby's—a fragile structure with the
weight of the light headset resting on it not well distributed. There
had to be a solution.[1]

Mike was experienced in finding solutions, especially in new ways.
He had been fascinated with planes, boats, and machines since childhood. At nineteen Mike earned his private pilot's license, a childhood
dream, and has been piloting his own planes ever since. He developed
design skills handed down through the generations: around 1914 his
maternal grandfather, Ronald Young, had designed and built the first
automobile to appear on the Kitsap Peninsula in Washington State.
His father, Jim Dennis, designed hearing aids, aviation-helmet ear
seals and ear cups, and communications equipment. Mike was always
a tinkerer himself. People would bring him things and ask Mike if
he could figure out how to fix them or make this or that. He made
things ranging from fancy hardwood flooring and sporting goods
equipment to machines for packaging razor-blade components. When
they were married, he told his wife that tinkering was his passion.
They were stuck with whatever he could produce out of his garage.

Mike first thought about using sheepskin to soften the headset for his wife. He imagined the biggest footprint he could make for the top of her head and what shape would be reasonable. What popped into his mind was a picture of Audrey Hepburn wearing a funny little hat about the size of one's hand on her head. So he cut out a piece of leather and glued some wool to it. He glued the material onto the headset, and it worked. The headache problem disappeared.

Not long after his new discovery, Mike and his brother flew to an air race in California. While sitting around waiting for the clouds to lift, someone came by, lifted the headset out of the plane, and asked Mike what it was. Mike explained that he had made it to alleviate his wife's headaches when they flew together. The guy offered Mike fifty dollars for it, but Mike said no because his wife would have been very unhappy if he had sold it. Other people also wanted to buy it. Mike continued to say no, that it was a pain to make. Undaunted, one man said that Mike owed it to humanity to make more of them.

By the time he returned home, Mike had talked himself into a way to do it. He was going to the fly-ins anyway, and he could make a couple of dozen, take them along, sell them, and make some gas money. His wife reminded him that he had said no more starting businesses, but he assured her that it would be like a hobby. Mike had started out as an airplane mechanic and tool-and-die maker but found that these trades were not for him. He had owned a couple of businesses previously but had decided that he was not going to do that again. Eventually, Mike produced a number of the headsets that he took to fly-ins. They sold briskly. Soon people started calling him, wanting more. This prompted Mike to ruin a perfectly good hobby and once again start a business using the money from the headset sales. He named the new endeavor Oregon Aero. It was a real gamble since he had only $400 to start the company and three teenagers to support.

Initially, Mike concentrated on headsets. The Oregon National Guard took notice of Oregon Aero's headsets and asked him if he could come up with a fix for its helicopter pilots' ill-fitting, uncomfortable helmets. He accepted and made his foam to fit the helmets. Then the Guard wanted to know if Mike could fix the seats in their c-130s, so he started making seat cushions for them and then for f-15 pilots. He also developed a padded ejection seat that worked well with fighter aircraft at Brooks Air Force Base in San Antonio.

Mike was invited to Warner-Robbins Air Force Base, Georgia, and met with an Air Force colonel who was the assistant program manager for the Air Force's c-130 aircraft. Their meeting resulted in an agreement for Oregon Aero to provide the Air Force with the cushioned seats. The colonel later announced to Mike that they loved what he had done with the c-130 seats. When Mike expressed concern about the informal methods of procuring from his business, the colonel told him, "If you go through the front door, you will never get there." He was right; it took Mike a long time to get anything done going through the front door. However, it started what became a twenty-plus-year relationship with the Air Force for Oregon Aero. Almost every aircraft in the Air Force now is equipped with seat cushions made by Oregon Aero.

Then in the late 1990s, U.S. Special Operations Command (ussoc) took notice of Oregon Aero and began knocking at its door for help with the newly designed combat helmet that it called the mich. The new design of the helmet was a product of the battle fought in Mogadishu, Somalia, in early 1993 between U.S. military units, including Army Rangers, Delta Force, and members of Navy seal Team Six, and Somalian rebels, with the aim of capturing the warlord Mohamed Farra Aidid. The battle was later reenacted in the popular movie *Black Hawk Down*. In that battle problems with the heavy, unstable, and uncomfortable pasgt helmet became evident to the Special Forces operatives who needed a lighter, comfortable, and more agile helmet.

In that battle, some of the Special Forces soldiers replaced their pasgt helmet with a type of helmet known as a canoe helmet. The canoe helmet, a plastic recreational helmet used for activities such as cycling, skiing, and kayaking, was purchased off the shelf commercially and adapted for their use. Its unique feature was a type of padding inserted in the interior of the helmet providing a stable platform and comfort for wearers, allowing them to do their job without pain and interference caused by an unstable helmet. The canoe helmet didn't stop bullets, but it did stop fragments. It also did not have a duckbill, like the PASGT, which interfered with communication equipment and night-vision goggles.

Later, in 1997 two of the Special Forces operatives who had fought in Mogadishu and used the canoe helmet attended a helmet exhibit in Boston to look for a better helmet to replace the PASGT. They eventually met an official from the French-owned helmet manufacturer CGF (Casgues Gallet Francaise) Gallet who took a special interest in their quest. The CGF Gallet official returned to his company and sold management on the idea of designing a new helmet for Special Forces use. The company used the best design of commercial off-the-shelf helmets and came up with the Modular Integrated Communications Helmet, known by the acronym MICH. It became the first combat helmet to use an interior foam-pad suspension system and an advanced type of Kevlar that offered increased protection against handgun rounds.

The Army Special Forces Command (SFC) sent Mike the new helmet. It wanted the helmet to be stable and comfortable; otherwise no one would wear it. Mike saw that the helmet included polystyrene padding much like that used in motorcycle helmets. It also used nylon webbing and a nylon-mesh net like a hairnet. It involved three different, very incompatible concepts for a helmet all put into the same helmet. Mike believed somebody had thrown everything but the kitchen sink into the helmet trying to accomplish something but had ended up just creating many problems with stability and comfort.

The polystyrene alone can cause a transmitted shock of 300 g's, a fatal impact. The SFC told Mike that its goal was to make the helmet stable, comfortable, with a better shock absorber, to pass an impact test at 150 g's. Being straightforward, Mike let the SFC know what he thought of the inside of the helmet: "Well there's nothing wrong with this thing that ripping out the guts, throwing it away and starting over won't hurt." Normally, that kind of statement would have ended the conversation right there. Instead, the SFC's reply was to go ahead and tear it out. The SFC didn't care how it was done, which was unusual for a military agency. Additionally, it wanted the pads to be waterproof. According to Mike waterproofing is always difficult when considering how the anatomy is going to respond. The

SFC wanted it to go to sixty-six feet in salt water, stay for twelve hours, and come out dry. Mike asked why. "None of your business," was the reply. Mike thought, "Okay, so you are coming out of a submarine with a bunch of SEALS. Got it."

Mike knew that engineering the right foam would be very difficult, but he was determined to make it work right. Comfort is difficult to define. Mike gave it a lot of thought and came to the simple conclusion that it was achieved when the person wearing the helmet no longer wanted to take it off after long periods of time. After going through hundreds of thousands of dollars of failed experiments, Oregon Aero engineers arrived at an engineered two-layer foam that met the requirements set by SFC. The foam was made of a type of viscoelastic material and a high-rate wicking material that collects perspiration and carries it to the outside edge, where it evaporates to provide cooling and drying.

In explaining the science behind the development of his foam during an interview, Mike was animated with palpable passion and excitement about his efforts. He had discovered that two forms of foam were needed for total comfort—one for stiffness, which he colored blue, and one for malleability, which he colored pink. Both forms of foam behave differently at any given temperature but have the same characteristics with the ability to absorb an accelerated load. The point, Mike explained, "in having one layer slightly stiffer than the other layer is that one half of it is facing out against the helmet shell, where you want stiffer material that pushes back a little bit harder when there is rapid acceleration, and you want the softer material facing human tissue, where it moves more readily for the comfort." Unfortunately, Mike said, "The Army never listens to this. They think it's all hocus-pocus."

Further, he noted that "Oregon Aero pads are temperature sensitive and rate sensitive. Temperature means that when you put the helmet on, the temperature of your head causes it to reform itself and to take your shape." Pads should also have a low rebounding nature so if one pushes down on the pad with a finger, it rebounds slower than the finger. According to Mike "it [the pad] has a no spring-like nature. It's not pushing back. It has very little. It also will not cut off blood flow, won't cause ischemia—capillary blood loss, which is why the guys take their helmets off because their head hurts and feels like it's on fire because the circulation is being compromised."

"Springs are amplifiers," Mike said. As an example, "if one drives a car with no shocks, just springs, that driver would find that even on a smooth surface, the car would leap in the air because it lacked a shock absorber to deal with the road surface. The spring was there to isolate the driver from the activity of the road surface. The shock absorber is there to protect the driver from the over excitement from the spring because the spring would just keep amplifying."

Further, "when soldiers fall out of trucks, bullets strike helmets, or a blast wave or shrapnel hits, things happen fast. A blast wave peak g lasts two milliseconds. One can withstand some ferocious impact for two milliseconds. An airplane crash lasts fifty milliseconds. That is a lifetime by comparison. Fifty milliseconds is how long it takes to blink an eye."

"What you need in a blast event," Mike continued, "or even if a person bangs his head against a tank gun, is time to start accelerating the body in an opposite direction before the head experiences a critical acceleration to the brain. The head and the body are not separated. It's a homogenous event."

When a blast wave plows into a soldier, Mike explained, the helmet should provide sufficient protection to delay the onset of the shock wave. The delay is important to mitigate the effect of the shock wave until the entire body starts to move; then time and distance helps in the prevention of an injury. The foam should be engineered to buy the soldier a significant piece of time during a traumatic event. That something, Mike revealed, "is molecular deformation, a controlled deformation, molecule by molecule, of a material that, when it experiences an acceleration, its structure locks up, it doesn't want to be accelerated, and for a millisecond, it becomes a solid."

If a soldier is wearing a helmet with just a webbed, or sling, suspension system, Mike explained, the helmet shell will accelerate and smack the soldier at about the same time the shock wave does. "If the soldier has a padded suspension system in which the pads are too stiff . . . then the shock wave is going to transfer through and flex the skull before the head even starts to move. The soldier starts to move, but it's too late. The damage has already been done to the head. The acceleration to the head has already been done. In a correct padded system, the acceleration is trapped in the soft and dense bonded foam material and has trouble getting through. Once the soldier begins to move, the blast shock-wave event is either successful or unsuccessful. If the blast wave hits the helmet and penetrates through to the head, the brain will flex and become damaged."

Once the foam was engineered, the only thing left was the coating on the outside. The foam was covered with a high-rate, wicking fabric on one side to cool and dry the pad. On the other side was a Velcro hook–compatible fabric for attachment to the inside of the helmet. One could put the cover on and take it off to clean it and attend to the perspiration. Little did Mike know at the time that his tinkering with the foam would be a pioneering effort that would revolutionize the design of the combat helmet.

Once the foam-padding system was completed, the SFC decided it was perfect as specified for its MICH. The pads met the 150 g peak head form acceleration limit (the magnitude of acceleration that causes brain injury). The 150 g threshold was established as compatible with maintaining consciousness and enabling injured troops to stay in the fight. Although there were no existing helmet blunt-impact standards, the SFC ignored protocol because it was trying to develop a good helmet that would work in combat.

Oregon Aero's pads were initially procured directly by the Special Forces Command, bypassing the Army's primary equipment research center at Natick. Fortunately, the SFC had a unique priority within the Army to develop and purchase its own equipment. The development of the MICH began in 1997 as part of the SFC Personal Equipment Advanced Requirements program located at the USSOC in Tampa, Florida. Deployment into the field had begun by 2002, when it was provided to Army Rangers, Navy SEALs, and Air Force Special Operations. The helmet was also fully fielded to the Marine Corps reconnaissance community. In 2002 the U.S. Army Material Command named the MICH one of the 10 Best Inventions.

In the end the proof was in the helmet's actual combat use. After two years of fighting in Afghanistan, the Special Operations Forces troops using the MICH with Oregon Aero's padded suspension system did not report any significant brain injuries. In contrast the Marine Corps stated in a paper that, on average, its troops were experiencing ten brain injuries per day for two years with their old sling suspension PASGT helmet.

Soon the Army helmet czar, George Schultheiss at Natick, took notice of the pads being used in the MICH. Taking a cue from the MICH, Natick developed a new combat helmet to replace the PASGT called the Advanced Combat Helmet, or ACH, which had Oregon Aero's pad system installed. Schultheiss received a meritorious award for having developed the padding even though he actually didn't develop it. Mike didn't care. Schultheiss wasn't hostile to Mike at that time since Oregon Aero was making him look good. That relationship would later change.

Mike recalled he and some of his engineers "sat around one day and figured out that the Army was so fixated on the numbers that they were going to stumble onto an aluminum foam and they are going to look at the numbers for an aluminum foam and say that's the stuff to put into helmets. Two days later, we got a memo from the Army that actually said that aluminum foam looks like a good idea for padding in a helmet."

Also, Mike said, "if you are an engineer and you are looking for material that you can take off the shelf that has the most infinitely successful deceleration character, and all you are looking at are numbers that will stop shock but it will also kill you. You can take reticulated aluminum material, make it crushable, take the shape of your head, and can use it, but it would cause a soldier to bleed out. It would drive all the soft tissue through your head. It would look like your head went through a cheese grater. The point is you would never put the helmet on. So, why would the Army think about putting that inside of helmets? But, people have criticized Oregon Aero pads as being too soft." Yes, "It's too soft," Mike remarked, "but what they are not right about is what this can do when you [impact] it very fast." To illustrate his point, Mike struck a pad hard on a table. He hit it very hard but did not harm his hand, arm, or shoulder because of the foam's molecular structure. "Even if you hit it with the fingertip, you can't go through it. It dissipates the energy sideways. We are talking about billions and billions of molecules that make up the foam and that is how they behave."

In addition to his core company in Scappoose, Mike owns half of another company called Oregon Ballistics Labs (OBL) located in Salem, Oregon. OBL is a forty-thousand-square-foot facility spread out in eight buildings that suspiciously resemble the mini storage complex it used to be. However, today it's considered one of the most sophisticated ballistics test laboratory in the United States outside the government. The facility has a 15 m gun tube and a 30 mm cannon that is fired regularly. It also has a 5 m gun tube with a Thompson test gun in it that fires up to a .50-caliber bullet

OBL does almost all the production quality testing for the ballistics steel mills in the United States, for ceramics, and for DuPont's Kevlar, and it is a certified facility for testing body armor. In addition OBL conducts blast tests on combat helmets. The tests are conducted in a remote location near Bend, Oregon. OBL was testing helmets with its pads in them in 2005, before anybody was talking about it. The company did the research on its own, and no one asked it or paid it to do so. The purpose of the blast tests was to establish a baseline for the helmet with pads because OBL couldn't get any information from the military. Mike believed that in 2005 the Army had not conducted its own testing on blast wave effects and didn't in fact know what a blast test looked like.

CHAPTER 7

Mr. Helmet

Everything that I've wanted to do to try and help the men and women in the field I've had to do outside the normal Pentagon bureaucracy. . . . A combination of stringent performance goals, inflexible rules and undetected "culture creep" all conspire to leave behind a rigid organizational structure that can't switch directions when crises erupt. As a result, leaders have to go around the very bureaucracy that is supposed to help them, but turns out to get in their way.

—Former secretary of defense Robert Gates

Within the vast military complex are numerous small bureaucracies that have become very powerful entities within the sphere of their jurisdictions. Two such bureaucracies control combat equipment for the U.S. Army and the Marine Corps—equipment items vital to soldiers and marines in combat such as clothing, weapons, helmets, and armor. Within these bureaucracies, manned both by civilian employees (the permanent bureaucrats who dominate the bureaucracy) and military members (transient soldiers and marines who sometimes become bureaucrats in uniform by rotating in and out every few years), there is typically one individual (usually one of the civilian employees or retired military who return as civilian employees) who can be called a "czar."

The czar is the dominant, or alpha, bureaucrat who reigns over a specific type of equipment, sets its requirements, and influences the award of production contracts. In a position of power, the czar usually acts in a seemingly autonomous manner within the commonly robotic bureaucracy. He or she makes decisions within the bureaucracy, especially in relationships with production contractors and vendors, that sometimes end up being detrimental to the immediate needs of the soldiers in combat with rifles, armored vehicles, body armor, boots, helmets, and other items of huge importance. These decisions can also compromise the battlefield effectiveness of the equipment.

The Army's Natick Soldier Systems Center in Natick, Massachusetts, was for decades the location of the Army's helmet bureaucracy. The center was built in 1952 with the mission of researching, developing, and testing military matériel and equipment. This was where combat helmets for the Army were designed and developed and their requirements set. The center dominated the procurement of helmets and historically influenced which vendors were awarded contracts for manufacturing them. At Natick the czar was an engineer named George Schultheiss—known by his friends as "Sumo." Many in the helmet industry referred to him as "Mr. Helmet" for his development and position as program manager for the PASGT helmet. It was his baby, and he had bureaucratic ownership of it. He dominated the helmet business for a number of years. In fact he dominated his post so completely that he let helmet vendors know that if they wanted to be in the helmet business, they needed to have a good relationship with him. Unfortunately, this attitude would prove to be detrimental for the soldiers who relied on the helmet in Iraq and Afghanistan. A native of New York, Schultheiss graduated from Worcester Polytechnic Institute with a mechanical engineering degree in 1983 and went to work at Natick in the Ballistic Protection Department, where he became involved in the engineering and designing of protective gear.

The PASGT helmet was initially developed to overcome the many major deficiencies of the old M-1 helmet, known affectionately as the "steel pot," such as weight, stability, fit, and comfort. It was also to provide at least equal ballistic protection but at a lesser weight. To achieve the desired weight and protection performance, Schultheiss and the developers at Natick decided on a change of ballistic material from steel to Kevlar. Natick was given the responsibility of developing the helmet. Unfortunately, their efforts produced a helmet, despite its improved weight compared with the M-1, that did little to improve protection, comfort, and stability.

Pierre Sprey is a former official in the office of the assistant secretary of defense for systems analysis under Secretary of Defense Robert McNamara. Born in France, he entered Yale at the age of fifteen and graduated with an aeronautical engineering degree. After earning his master's degree in mathematical statistics from Cornell, he worked for Grumman Aircraft. Sprey was recruited to the Pentagon as one of McNamara's renowned "whiz kids" in the 1960s. The whiz kids were a group of experts recruited to turn around management of the DOD. Sprey advised McNamara on major areas of the military budget and worked on tactical air analysis for NATO. He was later one of the main designers of the F-16 and A-10 fighter aircraft and the protocol of testing for major weapon systems. While working on the design and testing of aircraft, particularly of close-support aircraft, he became more deeply involved with ground weapons and realized that ground warfare was vastly more important than air warfare. After he left the DOD, this interest led him to finding ways of reducing troop casualties in Iraq and Afghanistan, especially brain injuries.

With his focus on troop casualties, Sprey became outspoken on the subject of combat helmets generally and the PASGT in particular. "The PASGT was a terrible piece of work, Sprey stated.

> It was supposed to be lighter. Originally, they were trying to save a pound and a half on the helmet; instead they went back to the old weight of the steel pot, if not a little heavier. Two important things they ignored were tactical use, in that you cannot have a helmet that blinds and deafens you in combat. The helmet had much too much coverage of the ears that a soldier could not hear properly that could be disastrous and get people killed. It also had too much coverage overhead. You couldn't even fire properly if you were lying prone. The design and requirements were a bad job.[1]

"Despite the fact it was horribly uncomfortable, they continued the old tradition of a strap suspension system, leather straps around the soldier's head," Sprey added. "With the weight involved, it was very painful to wear, particularly in hot climates with sweat dripping down a soldier's face. It was also unstable. All of this is of enormous importance in helmets because when the troops are unhappy with helmets that are causing them headaches and stuff or it causes too much sweat; they are constantly taking it off to take a break from the helmet. When the helmet is off, it isn't doing much good. Comfort is central to the design of the helmet. A helmet does not work if it is uncomfortable."[2]

The PASGT, as with many important pieces of combat equipment, did not meet the needs of soldiers in combat and demonstrated the historical problem of a disconnect between the needs of soldiers and the parochial interests of the military-equipment bureaucracies that decided on its design. The Special Forces' MICH was much smaller than the PASGT, with a vastly improved design, largely because Special Forces bypassed the Natick helmet bureaucracy for its development. They had a small development board with heavy representation of Special Forces soldiers who knew exactly what they were after. So working together with helmet manufacturer, CGF Gallet, they got a great helmet.

The MICH was later put to the test in a major firefight in Afghanistan in March 2002, known as the battle of Robert's Ridge. The helmet performed well, allowing most of the soldiers to continue the fight for more than eighteen hours and survive. A MICH used by a survivor of the Robert's Ridge battle was brought back to Natick as evidence that the helmet worked. It wouldn't be long before the rest of the Army wanted a similar helmet with Oregon Aero pad technology.

Natick's Mr. Helmet, George Schultheiss, took notice of the new MICH. Some industry observers felt at the time that he was not happy with the development of the MICH as it would have been a potential threat to his PASGT program. However, he soon embraced the MICH and its Oregon Aero foam-padded suspension system, which subsequently led to the development of the ACH that initially included Oregon Aero pads for regular Army soldiers.[3]

CHAPTER 8

Shut Out

Oregon Aero filed a patent application in August 2001 for its highly engineered foam pads titled, "Body-contact Cushioning Interface Structure." Earlier George Schultheiss had sent Mike Dennis a green foam and suggested that Oregon Aero make new pads out of it. To Mike the green foam looked like the best stuff, but it was not for human interface. It would work fine for an ejection seat but was not good for helmets because it was too stiff. Mike noticed that Schultheiss had dipped it in a Plasti Dip substance like a liquid vinyl. Mike said that Oregon Aero sprays a coating over its foam.

Once Schultheiss heard of Oregon Aero's patent, he was livid. He called Mike Dennis and actually accused Dennis of stealing his patent. Skeptical, Mike asked him if he had a patent. Schultheiss could only reply, "Well, you stole my idea." Mike told him, "George, the only similarity is there is some coating outside and foam of some kind on the inside. Our patent isn't about putting goop on the outside of the foam." Mike couldn't understand why Schultheiss was so upset. He told him on the phone, "The thing you sent was a piece of foam dipped in vinyl; that's not what we make here. If you like, I would be happy to buy you a plane ticket and have you come out, and I would take you to the factory where we make it and show you how it's done."[1]

According to Mike, Schultheiss arrogantly told him he knew all there was to know about making foam so he didn't need to travel to Oregon Aero. When Schultheiss continued to claim that he had invented the new foam, Mike became worried that he had an angry guy on his hands. Schultheiss's call sounded like it spelled trouble for Oregon Aero, as a supplier, and he feared his retaliation.

Mike didn't realize another deep-seated reason for Schultheiss's rejection of his offer to travel to Oregon Aero. Bureaucrats, within their workplace, assume the identity of the bureaucracy and function well within that environment. If the bureaucrat steps out of his environment and into the private competitive world, he or she may fear becoming impotent, devoid of power, like a fish out of water. The bureaucrat is a master of his environment but can feel jealous and overwhelmed outside it.

In 2002 Natick had introduced the Advanced Combat Helmet (ACH) for ground troops. It was adopted from the MICH and developed by helmet manufacturer CGF Gallet. According to a source with inside knowledge of helmet development, there was some resistance within Natick to development of the ACH because of its threat to the PASGT program. However, pressure from regular Army soldiers, as they discovered the benefits of the MICH, increased significantly and contributed to Natick's decision to develop the new helmet to replace the unstable and uncomfortable PASGT.

Oregon Aero originally provided the seven-pad kit for use in the ACH through its contract with CGF Gallet and later MSA, the company that bought out CGF Gallet. These pad kits were used in the initial run of the six hundred thousand ACHs that began fielding in 2003 to replace the PASGT. However, Oregon Aero's position as the primary provider for the padded suspension system in the ACH would soon deteriorate as conflicts developed between Schultheiss and the company.

Out of about six hundred thousand Oregon Aero pad kits delivered for the ACH, only one hundred thousand were installed in the helmets. Then, in late 2003 and early 2004, Schultheiss began to be concerned about the "high costs" of Oregon Aero pads. Mike thought Mr. Helmet had developed a personality conflict with one of the Oregon Aero sales representatives. Sources close to the situation believe that Schultheiss wanted to remove Oregon Aero as the pad vendor for the ACH and task the National Industries for the Blind (NIB), a nonprofit operation under a federal purchasing program, as the primary purchaser for the pads. Helmet pads were identified as appropriate for production by the blind in accordance with the Javits-Wagner-

O'Day Act of 1938 as amended in 1971, under the leadership of Senator Jacob Javits, to provide employment opportunities for people with severe disabilities, allowing them to sell products to the federal government.

Oregon Aero's contract with MSA, the main manufacturer of the ACH, was tied to the volume of ACHS produced. However, in 2004 MSA could not agree on a contract extension with Oregon Aero and stopped placing orders with it for the ACH. Schultheiss now had a clear road to eliminate Oregon Aero from providing pad kits for the ACH. Despite the cutoff from MSA on the ACH, Oregon Aero continued to fill orders for the MICH also being supplied by MSA. Up to that point, it had a track record of supplying the military with products such as seat cushions and ejection seats for aircraft since 1990. By the time it started pad development for the MICHS, Oregon Aero was selling $14 million of products a year to the military.

Rumors started swirling in the industry that Natick was going to reveal a better pad kit product to compete with Oregon Aero at a cheaper price. Mike Dennis's response was, "Cool, if something better comes out, you buy it right away because this is about keeping guys alive. It isn't about us at all. It has never been about Oregon Aero." But was it a better product? Mike was concerned. He wondered why Natick, or the Army in general, wanted a cheaper pad. Mike was getting numerous emails from soldiers and family members of soldiers in Afghanistan and Iraq thanking his company for keeping them alive after bad things happened. He could not imagine that the system was going to turn their noses up at it and come out with something that could injure people. It was incomprehensible to him.

By late 2004 Natick began using the NIB as both the purchasing agency for the foam pads and as a participant in the manufacturing process. The NIB, in turn, chose Team Wendy (TW) as its subcontractor. Critics raised the question if the handoff of the pad kit procurement to the NIB preempted the responsibility of the contracting officer to make an unbiased selection. The NIB justified its selection, stating that Team Wendy provided the "best value for their organization and the Army and . . . the best non-ballistic protection." With its selection Team Wendy, in a press release, announced that it was the leading provider of helmet pads for the Army. The Army labeled this arrangement as an NIB–TW "partnership." Once the NIB received the pad components from TW, it cut and assembled the pad sets for distribution. The completed pads sets were then sent to the Defense Logistics Agency (DLA) depots, to Natick, or to helmet shell manufacturers.

It was during the course of this drama in 2004 between Natick, TW, the NIB, and Oregon Aero that Operation Helmet started sending Oregon Aero pad kits to Army soldiers and marines. Doc Bob didn't become aware of the interjection of the NIB as a purchasing layer for the Army until a couple of years later. He learned that the NIB, besides purchasing the pads for the Army, also installed and stapled them into a bag. Doc Bob believed that the NIB was a type of self-perpetuating dynasty that could pick and choose who was going to give it the best piece of the action. According to Doc Bob, "the 'partnership' between TW and the NIB is similar to insider trading; the NIB is congressionally mandated to be included in the procurement and TW made them the best offer. Money talks."[2]

CHAPTER 9

Cost More Important Than the Troops?

Concern for man and his fate must always form the chief interest of all technical endeavors. Never forget this in the midst of your diagrams and equations.

—Albert Einstein

In the realm of helmet-impact research, 300 g's is considered an impact to the head that is fatal. To simplify the higher the g's, the more severe the impact on the head. In 1971 the Helmet Impact Criteria, developed by the Army, specified that peak head acceleration should not exceed 400 g's at 17.1 feet per second. This standard was used in the Army's aviator helmet at the time. Later research in 1980 at the Fort Rucker U.S. Army Aeromedical Research Laboratory (USAARL) disclosed that peak head acceleration far less than 400 g's produced concussive injuries to the head, leaving the Army aircrew members incapacitated following a crash. Improvements were eventually made to aviator helmets to reduce velocity forces, but by the late 1990s, with the development of the Special Forces' MICH, peak head impact was established at 150 g's. Specifications for the Army's ACH continued using the level of 150 g's as the standard—that is until the fall of 2005.

In the fall of 2005, Natick inexplicability reduced the peak head impact requirement for the ACH to 300 g's even though it was known to be fatal to the soldier. It came as a shock to Oregon Aero's Mike Dennis. He thought it was a typographical error and that Natick had made a mistake. Why would Natick arbitrarily reduce protection of a soldier's head at a time when TBI was on the rise? The revised specification change was crafted in subtle and vague language stating that the average number of hits could exceed 150 g's and no single hit could exceed 300 g's. However, if a soldier was hit once at 300 g's, it would kill him. Mike's theory was that Natick's motivation was to make pads cheaper but that effort couldn't meet the 150 g level.

In January 2006, already in a strained relationship with Natick's Schultheiss over Oregon Aero's rejection as the primary vendor to the ACH for its padded suspension system, Mike Dennis requested a meeting with Schultheiss to address his concerns over the reduced impact standard, given that there was helmet technology that could perform significantly lower than 300 g's.

On February 8, 2006, it was a clear, cold winter morning, in Natick, Massachusetts, just outside Boston. The chief operating officer for Oregon Aero, Tony Erickson, accompanied by a contingent of other Oregon Aero officials including a technical representative versed in foam engineering, walked into a drab meeting room at the Army's Natick Soldier Systems Center. Waiting in the meeting room was George Schultheiss along with one of his assistants and a civilian contractor representing the Marine Corps named Michael Cordega.[1]

The meeting started out with some tension over Operation Helmet and Doc Bob. Cordega told the group that he believed Operation Helmet was Oregon Aero's "boy" that controlled Doc Bob's activities and that Doc Bob actually worked for Oregon Aero. Schultheiss and his assistants indicated their agreement. Erickson responded that Operation Helmet was not part of Oregon Aero but was just a customer, that Doc Bob didn't work for the company. Cordega and the others said they were glad Erickson had cleared that up.

However, Erickson had a specific and uncomfortable question for Cordega. Why did the Marine Corps's new LWH require much less protection than the Army's ACH? Erickson explained that the LWH, with its padless suspension system, amounted to a 36 percent reduction in protection that would likely allow head injuries almost twice as severe as the ACH allowed. Cordega could only answer that it was the decision of the Marine Corps.

Continuing along the same lines, Erickson asked Schultheiss why Natick had reduced protection of the ACH by raising the impact threshold from 150 g's to 300 g's, clearly moving backward in technology. Incredibly, Schultheiss claimed that no other pad manufacturer could compete at the 150 g's level and that he did not want to sole-source the pads but wanted to expand competition. By lowering the standard, he believed more companies could compete. Erickson was very upset with Schultheiss's push to favor competition at the expense of head protection for the soldier. It was especially egregious during wartime, when IED blasts with the resultant incidents of TBI were dramatically increasing.

Erickson wasn't experienced dealing with military decision makers. He naively thought there would be a logical, reasonable explanation for the standard being raised to 300 g's—that it was simply a mistake, an error, that would be fixed right away, or that Schultheiss had found a cheaper product that worked better—not an explanation that made no common sense. Erickson was hoping for a plausible explanation. He had gotten wind that Schultheiss was hunting for a replacement pad but was very suspicious of changing the spec to be allowed to make that purchase.

Then inexplicably Schultheiss praised Oregon Aero pads. He claimed that he was proud of its pads and that Oregon Aero rose to the occasion by meeting all the deliveries on time in support of the war effort. He called the Oregon Aero pads the "Cadillac of pads." Then Schultheiss dropped the other shoe, that Oregon Aero pads were too expensive for the Army. He claimed that they were fine for the Special Forces *but said the Army was not going to spend as much money on the average troops.* He explained that the military had invested a lot of money in Special Forces soldiers, who take better care of the equipment and, frankly, are smarter. On the other hand, Schultheiss told the group that the regular soldiers did not take care of their equipment and less was invested in them, so the Army was not going to spend as much money on them.

Erickson recalls that Schultheiss said, "The Army doesn't feel they are worth it," not that "he" didn't feel they were worth it. The Army wouldn't spend the extra money on the regular soldier.

Erickson found Schultheiss's statement incredible: "He is an official representative of the Army in a position responsible for this piece of protective equipment. How could he make a statement like that?"

He was unprepared for such an explanation, one that he would never have anticipated. Erickson asked Schultheiss if "the Army was not willing to spend an extra twenty dollars to protect the regular soldier?" He was outraged that the Army did not think its soldiers were worth it and wished he had a tape recorder. He recalls thinking at the time that "if the mothers and fathers of the soldiers could hear him saying that they would lynch him at a tree outside."

This callous attitude was Schultheiss's and by extension the attitude of the helmet bureaucracy in developing and manufacturing combat helmets for regular soldiers—cheaper pads, less protection.[2]

Even more stunning news was revealed at the Natick meeting. Schultheiss, from across the table, held up a test report that Erickson had not seen or heard of before. Pointing to it he said it was an August 2005 draft USAARL test report on blunt-impact performance of the ACH and the paratrooper and infantry PASGT (Report 2005-12). Schultheiss told Erickson that according to the test results, Oregon Aero failed the performance tests by scoring close to 300 g's on hot temperature tests. (The USAARL conducted environmental tests on helmet pads at three different temperature levels: cold, ambient, and hot.)

Then Schultheiss ridiculously claimed that in order for Oregon Aero to compete, it had to raise the impact standard (to 300 g's) as a favor to the company. The company would be out of the competition at 150 g's. Erickson thought the Army should not change the standard to a level that could result in serious or fatal brain injury to a soldier just as a favor to anyone, based on some secret testing. It didn't make sense. He was shocked that the impact standard was supposedly changed for Oregon Aero based on a test the company couldn't examine or analyze.

Erickson was confounded and angry at this obviously flawed revelation. How could Oregon Aero have failed a performance test? The company had never in previous military and independent testing even come close to failing or even scoring close to 278 g's. This one-time test was 200 percent higher than dozens of independent lab results had shown for the company's pads. In fact those results were always around 100 g's. Oregon Aero had always, and continued, to meet the 150 g's standard required for the MICH. Schultheiss himself was part of the Natick Soldier Systems engineering department that had done its own tests in 2003 showing Oregon Aero pads passing an even more rigorous 150 g test with flying colors.

If Oregon Aero had failed a performance test six months ago, Erickson wondered why the company was not officially notified before the meeting. Why wasn't there even one email or letter from anyone in the DOD noting the company's failed performance tests? Why was there no correspondence claiming in 2005 that Oregon Aero no longer met its original helmet standard, which the company had achieved between 1999 and 2005? He suspected it was because no such correspondence ever existed.

It was obvious to Erickson that something was very wrong. Three rapid thoughts flew through his head: the company's pads had actually failed; the testing was flawed in some way; or more likely, it was not really Oregon Aero's product. He asked for a copy of the report, but Schultheiss said he was not allowed to have it because it was not available to the public.

Further suspicion of Schultheiss's test claim was aroused when New York–based independent lab Intertek conducted impact tests of Oregon Aero pads in February 2006, around the same time that Schultheiss was saying Oregon Aero scored a peak acceleration of 278 g's on the hot temperature tests. Instead, Intertek's hot temperature test disclosed a result of 78.6 g's peak acceleration, a major difference from the results of the alleged USAARL tests. Knowing his product well, Erickson fully believed the pads attributed to Oregon Aero by Schultheiss were not actually its product. It was the only conclusion that made sense to him.

USAARL eventually sent Oregon Aero a copy of its 2005-12 report, including the photos of the helmet with the pads that were tested. The photos immediately made clear two very important facts. First, the report did not claim that the pads tested were Oregon Aero pads. In fact, not one pad set was identified with a manufacturer. Second, the photos revealed that the pads used in the tested helmet were unmarked, and Oregon Aero never produced unmarked pads or delivered such pads to the Army.

From the evidence Oregon Aero could only conclude that the pads Schultheiss claimed were Oregon Aero pads were not its pads. Mike Dennis also knew this because whenever he sent pads to a laboratory for testing, he photographed them and either hand-carried them personally or hired a reliable courier to transport them to the lab. As would later be revealed, Natick supplied the so-called Oregon Aero pads to USAARL. They did not come from Oregon Aero.

In Erickson's later phone conversations with Schultheiss, he revealed that the helmet manufacturer was charging the Army $102 for the Oregon Aero pad sets, placing the blame on Oregon Aero for the high costs. However, Erickson explained that Oregon Aero charged the helmet manufacturer $62 for a pad set. Its cost to manufacture a pad set, unburdened, was about $37. Thus the helmet manufacturer was making a $40 profit per pad set off the Army. Schultheiss resorted to shouting at Erickson, saying the helmet manufacturer was telling him that the problem with the costs arose because Oregon Aero was charging too much. Team Wendy was charging only $38 a pad set. Erickson realized that the helmet manufacturer was almost doubling the price of the pads to the Army and blaming Oregon Aero for the high costs. Erickson believed that if he had known the numbers, he might have been able to rebut the accusation or at least talk about it with Schultheiss.

Oregon Aero found itself in a predicament and suspected it was brought on by the machinations of Natick, especially Schultheiss. The company was being blamed for the high costs of its pads to the Army, when it appeared a helmet manufacturer was the culprit by almost doubling the actual cost. In addition Schultheiss had ambushed Oregon Aero by claiming it had failed a performance test with numbers four times higher than the company had ever seen previously.

To Oregon Aero testing at body temperatures was the most important aspect of determining how the pads would react. According to Mike Dennis, the pads would take on the same temperature as the head in the contact area. Human bodies are dense thermal masses, the pad is a very ineffective thermal mass, and the body wins. The company never saw high temperature or low temperature as an issue.

Doc Bob also criticized the testing method for extreme temperatures used by USAARL. He reasoned that heating the pads and the head form to 130 degrees Fahrenheit or cooling them to 15 degrees were not realistic because the head acts as a heat sink causing the pads interacting with the head to assume body temperature. He wondered why anyone would test the headform at 130 degrees with the helmet on it. A human being whose head reached 130 degrees would be dead as he or she would be at 15 degrees.

When Pierre Sprey was apprised of Schultheiss's comments regarding the more expensive pads not being made available for the regular soldier, he was taken aback. "The fact that the soldier is not worth the cost is an appalling excuse," he said. "But it reveals a little bit of the civilian bureaucracy mentality."[3] Sprey explained,

> You could never get a real Army combat unit to say such a disgusting thing and besides which it may never be true. He would assume it was not true in that the average soldier is perfectly capable of dealing with helmet pads just like the Special Forces do. There is not a huge difference in intelligence. There may be some difference in focus and training. Of course, the life of the average soldier is worth just as much as the Special Forces soldier. You have to remember, these are the people who have been rumbling for years about the special privileges the Special Forces get allowing them to buy their own stuff and not having to obey the requirements of the Natick bureaucracy. All that is being reflected in what they do.[4]

Given his deep knowledge of military testing procedures, Sprey disagreed with Natick's approach to extreme temperature testing of helmet pads:

> They are testing at cold temperature, somewhere below freezing, at ambient temperature which is room temperature, and they are testing at 130 degrees Fahrenheit. This, in itself, is an astonishing piece of incompetence. This is crucial, the hot tests, the insistence of doing tests at 130 degrees which is ludicrous from the point of view of combat. If it gets to 130 degrees inside your helmet, I guarantee there is not a person in the world that would wear that helmet. It is ridiculous to test the pads like that. Same with the cold. If it's 10 degrees

below freezing inside your helmet, you're going to have frostbite of your ears or whatever part of your body is touching that temperature. It's not conceivable that this is a sensible way to test helmets.[5]

A good example of the disconnect between the hot-temperature test conducted by the USAARL and the reality of how pads work in a hot environment during combat was a comment to Doc Bob from a marine who had worn Oregon Aero pads on patrol in Iraq in 2005. The marine said that despite the hot temperatures of summer that reached up to 120–130 degrees, he never experienced his helmet pads getting hot against his skull.[6]

After the Natick meeting, Oregon Aero was not finished dealing with the problem of the hot-temperature test with the Army. A few months later, Mike sought a meeting with USAARL, Natick, and the Program Executive Office (PEO) Soldier at USAARL's site at Fort Rucker, Alabama, in an attempt to reconcile the test results of the 2005-12 report that purportedly involved Oregon Aero pads. PEO Soldier was organized in May 2002 with the purpose of integrating the procurement and fielding of equipment for soldiers with different capabilities. The meeting would be an opportunity for Oregon Aero to directly confront USAARL on its test data and the company's disqualification as a pad vendor based on that data.

The meeting, scheduled for May 2006, began falling apart when officials from Natick and PEO Soldier began making excuses for not attending the meeting on the scheduled day. It became clear to Mike that Natick and PEO Soldier were trying to avoid attending the meeting altogether. Both parties finally agreed to meet on August 9, 2006, but Natick and PEO Soldier subsequently announced that their representatives would not attend the meeting due to "a travel budget shortage."

Natick tried to cancel the meeting altogether, claiming that there was no point for a meeting since no one would be there. Not to be deterred, Oregon Aero insisted its representatives would attend anyway. USAARL officials said they would attend the meeting and discuss anything except ballistic helmet-pad testing, which was the whole point of the meeting.

On August 9 Oregon Aero's Tony Erickson, along with four other company representatives, showed up at USAARL offices. B. Joseph McEntire, the USAARL engineer who had authored the 2005-12 report, and other USAARL lab officials met with them. Right away McEntire announced that USAARL was prohibited by PEO Soldier from discussing the helmet test results. However, he was willing to discuss aviation helmet issues, acoustics tests, or aviation seating (seat cushion), topics, which Oregon Aero was not interested in discussing. It was Erickson's impression that USAARL was under a gag order not to discuss the 2005-12 report test results because it was damaging to it specifically and the Army generally.

Despite the gag order, the lab officials seemed to talk about the 2005-12 report without actually mentioning it—talking around it in code. Despite the test results, the lab officials spoke favorably of Oregon Aero pads and alluded to a systemic lab problem for testing helmets for blunt-impact resistance. In the end the testing issue with Oregon Aero was not resolved. Natick and PEO Soldier would continue to use the test results as a basis to disqualify Oregon Aero as the pad vendor for the ACH. According to Tony Erickson, he later met McEntire several times at different events where McEntire was unaccountably hostile to him and Oregon Aero.

CHAPTER 10

Search for Justice

With raw emotions lingering, Jeff Kenner found it difficult to handle the termination from Sioux Manufacturing. His wife was humiliated by the affair rumors—more
angry with Jeff than the company. He tried to explain to her that the
incident with Tammy Elshaug was completely innocent; it was just
a tawdry rumor contrived to embarrass both of them and gave the
company an excuse to fire them. In addition to his concern about saving his marriage, he also needed to think about his finances.[1]

Despite his concerns Jeff knew it was time to move forward and
make things right. He decided to find an attorney, one who would
represent both Tammy and him in a wrongful termination lawsuit
and blow the whistle on Sioux Manufacturing's Kevlar weave-shorting
problem. However, he was unsure how to go about it. As driven as
he was to achieve some sort of redress, Jeff thought it important to
discuss options with Tammy, since it would affect her as well. In the
meantime he secured a job with a friend hauling oil, propane, and
fuel for his company. It didn't pay anywhere near what he was getting at Sioux Manufacturing, but it was better than nothing.

If he was going to blow the whistle on Sioux Manufacturing, Jeff
thought he should try to retrieve records from the Quality Assurance
Department to prove that the company was shorting the weave. He
was afraid Sioux Manufacturing would destroy incriminating records.
However, retrieving those records would prove very difficult now that
neither he nor Tammy was employed at the company. Jeff recalled
there was a U.S. Department of Defense representative, known as a
quality assurance representative (QAR), who had an oversight office at
the Sioux Manufacturing plant. He decided to reach out to him for

help, to report the weaving problem, and try to talk him into making copies of records, especially the thread-count records.

The QAR worked for the Defense Contract Management Agency (DCMA), a component of the Department of Defense. As the name implies, DCMA administers DOD contracts under delegation from the military's program contracting officer (PCO). The PCO deploys QARS, who monitor contractors' product performance through onsite surveillance for military buyers, to ensure compliance with the terms and conditions of the contracts. Jeff called the QAR and asked him if he had inspected the Kevlar pattern sets for the PASGT helmet. He advised Jeff that he had not been delegated that responsibility. He explained that he only "inspect[ed] contracts from the U.S. Army Tank-Automotive and Armaments Command," contracts that provided Kevlar shields for armoring Humvees. Jeff did not recall ever seeing the QAR in the weaving department, but he had seen him in other areas of the plant.

With an opportunity at hand, Jeff got right to the point with the QAR. "The company was committing fraud against the government by shorting the weave on the pattern sets on the UNICOR contract," he explained to him. "I can't get involved in that," the QAR replied. "It's against agency policy for me to do that." He suggested that Jeff blow the whistle through the DOD fraud hotline, but he did not have the contact information. Without hesitation Jeff asked him if he could make copies of thread-count logs that would be filed in the Quality Assurance Department. He explained that he needed the records to prove the fraud. "No, I can't do that," the QAR told him. Jeff ended the call frustrated that the QAR, the guy who represented the DOD, would not look into the problem or obtain records for him.

Over the next few months, Jeff began calling attorneys but couldn't find one to take his case. The obstacle seemed to be tribal sovereignty, which gives Indian tribes authority to govern themselves within the borders of the United States and establish their own laws. He called as many as seventy-four attorneys throughout North Dakota, Minnesota, and South Dakota without success.

Jeff was beginning to fear that he was out of luck finding an attorney. Unable to sleep one night, he wandered into the living room, where he turned on the television. He was staring at the TV, not really noticing what was on, when he was jarred to attention by a commercial for a law firm in Hawaii that dealt with government fraud. An attorney talked about how his firm worked with whistleblowers suing government contractors who were defrauding the government. "Wow," Jeff thought. "God must be looking out for me." As the phone number crawled across the screen, Jeff fumbled for a pen and paper and scribbled it down. "I'm going to call them in the morning and see if they will take my case," he promised himself.

In the morning Jeff called the law firm in Hawaii. One of its attorneys got on the line and explained to him a federal law, called *qui tam*, used by whistleblowers to take action on behalf of the federal government against a contractor who was defrauding the government. Jeff was intrigued. He asked the attorney if his firm would take his case. The attorney said he was interested, but the firm was too busy and wouldn't be able to handle it.

Disappointed but fascinated with the possibilities of using the *qui tam* law for the weaving issue, Jeff searched the internet to find more information on this unique law. He learned that the legal term *qui tam* (*Black's Law Dictionary* pronunciation: kwày tæm) is derived from the Latin *qui tam pro domino rege quam pro sic ipso in hoc parte sequitur*, meaning "who as well for the king as for himself sues in this matter." It is a provision of the Federal False Claims Act that allows private citizens to file a lawsuit in the name of the U.S. government, charging fraud by government contractors and others who receive or use government funds, and to share in any money recovered.

Historically, private citizens as far back as the thirteenth century in England used *qui tam* actions to gain access to the royal courts. In the United States, *qui tam* has been around since 1778, although it was seldom successfully used until 1863. In 1863, at the urging of then president Lincoln, Congress enacted the Civil False Claims Act, including the *qui tam* provision, as a weapon to fight procurement fraud that was proliferating during the Civil War. Thus the *qui tam* law has also been known as the "Lincoln Law" and the "Informer's Act." The federal government watered it down during World War II during the massive war buildup.

However, in 1986 Congress was seriously concerned about procurement fraud, which was rampant during the giant buildup of defense spending during the 1980s due to inadequate efforts by federal law enforcement and procurement and oversight agencies to control it. Additionally, legal obstacles within the *qui tam* law made it difficult for whistleblowers and prosecutors to bring *qui tam* actions. As a result Congress passed amendments to the False Claims Act increasing the whistleblower's share of the recovery to a maximum of 30 percent, increasing the powers of relators (whistleblowers) in bringing *qui tam* lawsuits, increasing the damages and penalties that can be imposed on defendants, and removing many legal obstacles to make it easier to bring actions before the court, increasing the likelihood of success. As a result *qui tam* actions increased dramatically and have been one of the most effective and successful means of combating government procurement and program fraud and returning more than $48 billion to the U.S. Treasury.

With a better understanding of the *qui tam* law, Jeff was intrigued and excited about its possibilities, especially the means it offered to achieve redress for the military and the troops for the fraud perpetrated by Sioux Manufacturing. He felt a moral obligation to pursue the case and was compelled to proceed because it was the right thing to do. He called Tammy and conveyed his excitement about *qui tam* and its possibilities for pursuing a case against Sioux Manufacturing. Also intrigued she agreed to be involved in the effort. Her frustration with the wrongdoing and its consequences for the troops were paramount in her decision. Her father and brother-in-law served in the military, and she had a nephew who was in the Marines. Some of the fathers of her children's friends were in Iraq. Both of them agreed that it was time to go public.

Now focusing on *qui tam* attorneys, Jeff narrowed his search on the internet and landed on a website aptly named federalfraud.com. On March 27, 2006, now greatly motivated, he called the law firm associated with the website, Campanelli & Associates, located in Mineola, New York. Eric Eubanks, one of the Campanelli associates, took the call. Such calls at the law firm were routinely screened by a paralegal or an associate, and if the caller met certain criteria, the inquiry would then go to the main partner, Andrew Campanelli, for his review. Generally, Campanelli established criteria before he reviewed a case, including a minimum base claim of $15 million before trebling the damages and a solid relator.

Eubanks felt Jeff's call met the firm's criteria, and he brought it to the attention of Campanelli. Finding the information compelling, Campanelli returned Jeff's call right away. Having talked with a thousand potential relators, Campanelli had developed an ability to size one up for his or her potential and could get a sense of the sort of person who was making the inquiry. Almost from the outset, he could tell that Jeff was honest, forthright, and genuine. Jeff provided a summary of the case and mentioned Tammy Elshaug as his potential co-relator. Campanelli felt strongly enough about Jeff and his case that he immediately agreed to meet them in Devils Lake, something he did not ordinarily do with potential clients. Jeff and Tammy were excited that a lawyer from New York would take the time to travel to their remote area in North Dakota just to meet with them. They set April 4, 2006, for the meeting.

Not long after Jeff's termination, the weaving machines at Sioux Manufacturing started breaking down. It became a serious problem that cut into the company's profit margin. According to a friend and former coworker of Jeff's, management wanted to rehire him because he was the only one who could fix the machines. After hearing this information, Jeff thought about the possibility of getting rehired and having access to the records he needed to prove his case. First, however, he wanted to meet with the attorney.

On April 4 Jeff and Tammy met Campanelli in the conference room of the Comfort Inn, one of the few small budget motels on the Highway 2 strip that runs along the outskirts of Devils Lake. Both were nervous about the meeting, but Campanelli found that they were more laid back then he had initially expected—not in the normal tense and excited state of most of his new clients with potential *qui tam* actions. He was again struck by how genuine, straightforward, down-to-earth, and honest Jeff was and was elated that Tammy seemed to have the same values.

Not pulling any punches but in a direct and calm manner, Jeff and

Tammy explained to Campanelli what had transpired at Sioux Manufacturing. Campanelli was impressed with their compelling narrative of the case and their grasp of the evidence needed to prove it.
His legal instincts told him it was a very good case with the potential
of also being a very large one. He knew immediately that he wanted
to represent them. It was clear to him that they were not in it only
for the money but were genuinely concerned about doing the right
thing, which in his view, was the best type of relator to have. By the
end of the interview, he sensed a connection with Jeff and Tammy
because he firmly believed they were two of the most decent people
he had ever met. It was now time for Jeff and Tammy to decide if he
was going to represent them. "You don't know me even though we
spoke on the phone," Campanelli told them. "I'm going to leave the
room, and you guys just hang out in the conference room and discuss it among yourselves. If you want to sign a retainer agreement
now or wait to think about it, I'm comfortable either way, whatever
you want to do."

Feeling confident that they had found the right attorney, Jeff and
Tammy agreed to take the plunge and seize the opportunity to finally
reveal what Sioux Manufacturing was doing to the troops. It was a
heady but ominous decision for them—two ordinary people from
a small town in the heartland taking on one of the biggest companies in the area. After about twenty minutes, Campanelli came back
into the room to receive their answer. They told him they wanted
to go forward and file a *qui tam* case, with Campanelli as their attorney. Campanelli reminded them that they needed to be sure about
their decision since they lived in a small town. Both quickly nodded
that they were sure. They felt that it was the right thing to do and the
best opportunity for gaining redress. The fraud had to be exposed.

Although a financial bounty is awarded to the whistleblowers in
successful *qui tam* cases, it was not what motivated Jeff and Tammy.
Like many whistleblowers who file *qui tam* actions, doing what they
felt was the right course of action in revealing the fraud and hopefully cause a federal investigation to "make it right" were the important motivations. They had tried to fix the problem internally but were
rebuffed at every turn. Instead, they had suffered for their efforts by

being demoted, with false affair rumors spreading around the plant, and then losing their jobs, which caused them severe stress and strained their respective finances, reputations, and marriages.

After the retainer agreement was signed, Jeff mentioned to Campanelli that Sioux Manufacturing wished to rehire him. As whistleblowers Jeff and Tammy were important "insiders." They were at the center of the process with direct knowledge of the company's past actions. They had some evidence, but Jeff wanted more proof, which he could obtain if he went back to Sioux Manufacturing as an employee. Concerned that the company would try to destroy evidence, he wanted to make copies of weaving records. Campanelli said, "OK, as long as you are doing it for the right reasons and you are only securing copies, not the originals, for the specific purpose of preparing for the *qui tam* case; there is nothing improper about that."

Within a week after the meeting with Campanelli, Jeff drove to Sioux Manufacturing, where he met with Carl McKay. Playing dumb about his knowledge that they wanted to rehire him, he told McKay he had a job coming up with Goldings, but he would just be wasting his talents working there. McKay, whose friendly manner contrasted starkly with his demeanor at their last meeting, seemed to be open to the idea of Jeff coming back to the company but was noncommittal. He implied that he wanted to hire Jeff back but didn't make an offer. Jeff played it cool but was desperate to get rehired and gain access to the records that would back up his *qui tam* case. He didn't care if they gave him only a dollar an hour. He was encouraged by McKay's friendly manner and left the plant believing he had planted the seed for being rehired.

About a week later, Jeff drove back to the plant and again met with McKay. This time McKay asked him if he would be willing to he hired as a consultant. "Ya, that would be okay," Jeff replied, holding his breath, not wanting to appear too anxious. McKay asked him to wait a few minutes. He wanted to check to see what they paid for consultants. McKay returned and said, "Name a number between fifty and seventy-five dollars per hour." Jeff was taken aback. Considering the low wage he had made when employed there, he couldn't comprehend making that much. "I don't expect that much," Jeff said. "How about forty dollars an hour?" "Well, what were you making when you left here?" McKay asked. "It was fourteen dollars or something," Jeff replied. "Well, how about eighteen dollars?" McKay suggested. Jeff agreed because he just wanted to get in the door. McKay told him that he would have Elvis Thumb, a production supervisor, call him.

After a couple of days without hearing from Thumb, Jeff called him. Thumb told him he had been talking to McKay and invited Jeff to meet with him at the plant. Jeff met with Thumb in his office, where Thumb told him that the company wanted to hire him not as a consultant but rather as a full-time employee, but with lower pay. Jeff's plan was just to get back in the plant to look for documents, so he readily agreed to the offer.

After Jeff agreed to be rehired, he walked with Thumb to the weaving department, where they met with Clarence Leftbear, the manager of the department since Jeff had been terminated. Leftbear told Jeff that he would not have to do too much to the weaving machines to get them going. Walking around the machines, Jeff saw that they had been trashed. Only a few machines were even running. He just shook his head and walked out, now knowing how much they really needed him and realizing he had a big job to tackle.

On April 24, 2006, Jeff started work again at Sioux Manufacturing. He thought of secretly taping his conversations with some of the employees to get evidence that the weaving process was still not in compliance with the contract. He had learned from his research that North Dakota was a "one-party consent" state, which meant that only one of the individuals in a conversation being taped was legally needed to consent to the recording. Thus Jeff believed he had the legal right to tape anyone in the plant, and he knew that it would be one of the best ways to gather evidence.

Before starting work Jeff called Tammy and explained to her what he was going to do. He told her he needed to get the weaving inspection reports so they would have enough information. He knew where the evidence was in the weaving area. Tammy told him to get packing lists that had the certifications with them. She was very nervous about his venture at Sioux Manufacturing to collect records. She didn't know how he was going to accomplish his goal. Jeff was jumpy about taping and about the possibility of being caught, but he calmed down by asking himself, "What are they going to do to me?" He also had to focus on the main reason he was rehired: to fix the weaving machines. Concerned about how long he could pull off this subterfuge, Jeff gave himself two weeks to fix the machines and get the evidence he needed for the case.

On his first day of work, Jeff drove in early, around 5:00 a.m., a pattern he would repeat each day he worked there. There was no one in the offices that early, but the file cabinets weren't locked. The file cabinets he sought, in the Quality Assurance Department, lined the walls and contained weaving logs, purchase orders, and other records. Since he was able to access the file cabinets, he had time to go through the files, find what he wanted, make copies, and carefully replace the originals. Even during the day, he found time to get into the file cabinets. It wasn't unusual for him to do so in the normal course of his business, since he often needed to look for certain records. When he found records he wanted, he simply copied them in front of people. It was nerve-racking, but surprisingly no one paid attention to him. Once he had copies of the records, he furtively stuffed them into his pants or jacket and strolled out of the plant.

The whole experience made Jeff edgy because it was out of character for him to be that sneaky. He was so honest that he used to feel guilty just taking a company pen home. Taking copies of records weighed heavily on him. He reported periodically to Tammy on what he had purloined, and she would tell him what other records he should capture such as DuPont purchase orders for Kevlar.

It was about three days into his work before Jeff attempted to record a conversation. First, he met with the quality assurance manager Rhea Crane in her office. In recording their conversation, Jeff wanted to get her to admit that the company had been weaving below the required thirty-five by thirty-five yarns per inch. He explained to Crane that he was trying to locate past history on thread counts to use in determining how the machines had been running. Nervously, he began the conversation by getting right to the point that he did not think the thread count had been thirty-five by thirty-five in years past. Crane appeared tentative and defensive, but declared, "I wouldn't doubt it, but we are not going to do it now because I want them to run at thirty-five by thirty-five now. I don't care what happened back then. I can't fix that," Crane explained. "I just started last summer."

Crane continued the conversation in a looser and more talkative mode: "A half pick [a term used to designate thread count] is not going to be a big deal, but some of those were like thirty-two by thirty-four. They didn't think it made thirty-four and a half. When we used to make helmets, I'm sure our cloth was plus or minus a pick, and if we were certifying to that mil spec back then, we were in violation, we were cheating, we were lying. Well, I can't do that. I won't do it," she explained to Jeff. "I think it came from up front to save piles of money," she continued, "so, big deal, then don't take the contract if you can't produce the material the way it's supposed to be produced. I haven't become a good corporate liar yet. I'm working on it."

Crane didn't stop there. Jeff encouraged her to continue—and it was just a matter of triggering talking points and letting her talk. She expressed her concerns about pick counts on material that were not up to standard. "I cannot put my name on this here test report that we print out, I send out with every lot of material, and it's supposed to be signed by the QA manager," she asserted. "I'm not going to put my name on there knowing it has thirty-three or thirty-four or whatever the picks per inch. All that stuff I put in the hold cage. I went and talked to Carl [McKay] about that, and I said that is not the requirement, and I said we used to say plus or minus an inch, and I don't know where they came up with that." "What did he say about that?" Jeff asked. "He said he didn't know either," Crane replied. "Oh, I think he does," Jeff countered. Crane added, "I think he's the one that was in on it to begin with, but he won't tell me that."

Later during their conversation, Crane pointed out that the Kevlar cloth weight, after weaving, had to be 13.5 ounces per yard to 14.5 maximum and can't be less than that. Jeff agreed. "Otherwise," she continued, "they're controlling you sending them too light weight cloth. I thought of this many times since I realized that [sic] how much have we sent out underweight, under picks. Now, I can't do anything about what's already happened, so forget it. But, if we ever had someone get killed and they decided to investigate why, because they thought maybe the helmet wasn't any good and they pick that thing apart." She snapped her fingers. "Oof," Jeff reacted. "Oof is right. They can do the same thing with any of our panels we sent out too," said Crane. "They've saved millions of dollars over the years, I believe, with less picks, but that's all we knew," Jeff offered.

Inwardly, Jeff was excited and urging Crane on, but outwardly his demeanor was calm and laid back. Crane kept going:

> Then why don't they go back to UNICOR and say, hey, we've proved that this is good, can't we go with less picks, see it's a government contract. The government says, uh uh, you're going by this spec, so you're stuck with that spec and that's what you should produce. I'm not going to send a shirt with one less button then [sic] what its called out for. You are going to get what you paid for. So, I mean, that's where I stand. I can't do . . . I can't stand in any other position. If they don't like it, then they can get somebody else to fill this chair, because I won't put my name on it. I've let things go that have been a little [unintelligible] . . . but this, you know. You should have been done doing this a long time ago. Now, half a pick, I don't think is going to make a big difference, but let's get it right. Let's get it right. If we ever got audited, you know what they would do to us, shut us down and fine us big time. Probably never see another government contract.

Crane assured Jeff that the company was now complying with the thirty-five-by-thirty-five standard.

Jeff decided he had enough evidence and ended the conversation. He left Crane's office feeling empowered because he had obtained some crucial information that would help the case. Later that day Jeff ran into Crane in the QA Department near the scrap table. He asked her if they used to round up the pick count and just write thirty-five by thirty-five. "Sometimes they do; sometimes I do," she admitted.

As the days dragged on, just being at the plant began to wear Jeff down. He almost felt like crying at times—the scars of what the company had done to him in the past had not healed. It was difficult getting through even two weeks of work because it was hard to know who was truly his friend and whom to trust. Yet realizing he was working on a type of covert mission, he knew he had to stay long enough to get the needed information. It was nonetheless painful to be there. This undercover work was out of character for him, enough so that he felt constant guilt for being deceptive. To his relief there was no talk of his rumored affair with Tammy during the two weeks he was there. Generally, the workers at the plant were happy to see him, especially since he could make the weaving machines hum along and get production moving forward.

One day Jeff thought he had lost his recorder. As he usually did, he patted his shirt pocket to assure himself that the recorder was there, but it was not. Panicked, he backtracked around the plant looking for it without success. If someone in the plant found it, his mission would be over. He would have a lot of difficult explaining to do. He realized he had to maintain a cool demeanor while looking for the recorder. When he didn't find it, he called Tammy and told her he thought he had lost the recorder. She became alarmed for his safety. At the end of his shift, he left the plant and walked to his pickup truck. He noticed his jacket, which he had left in the truck that morning. He picked it up and discovered the recorder in one of the jacket's pockets. Breathing a sigh of relief, he realized that his time at Sioux Manufacturing was short. It was all too risky. He had to quit soon.

Jeff didn't spend all his time gathering evidence. He also worked on the broken-down weaving machines to get them running again. Working with the machines gave Jeff the opportunity to record a conversation with the weaving manager, Clarence Leftbear. Jeff asked him how many years the company had been running low thread counts for the Kevlar pattern sets. "Probably since we started," Leftbear replied. Jeff told him, "I think Bill [Bill Cogburn, who was weaving manager before Jeff] was told though to cut down. . . . Cut down will save a lot of money." "Which they probably did," Leftbear said, "but I mean, whenever we would go through our inventory we would always have yarn left over." Jeff was pleased that he had gotten a good statement from a manager to back up his contention that the company's physical inventories always found excess Kevlar.

Before ending his short-time employment at Sioux Manufacturing, Jeff felt he needed to make a copy of his personnel file. When he got his hands on the file, he took a pencil and numbered the corner of each page in very small writing. He worried that the company could alter or destroy the file. The people at Sioux Manufacturing were not trustworthy, he thought, and they would lie or create false documents in the file, especially if he was successful at exposing their fraud. For extra backup he made a copy of the entire file and was able to take it with him.

In early May, after almost two weeks working at Sioux Manufacturing, Jeff decided it was time to quit. He believed he had obtained the information he needed to prove the allegations, and he couldn't stomach working there another week. Management thought Jeff was going to continue working at the plant indefinitely. He seemed to be happy and was crucial to maintaining the weaving machines. However, working there made Jeff ill with stress, and he had had enough. The records he had obtained and the candid conversations on tape would tell the story outside the walls of this little plant.

Before Jeff left for good, he needed to address the rumored affair that had led to his termination. He decided to ask Elvis Thumb about it, to get his take on the rumor. He walked into Thumb's office, asked him about the rumor, and recorded the conversation. Thumb told Jeff that he was aware of the incident when the firemen found Jeff and Tammy off the road. He admitted that the fire chief, Daniel Herman, was at the plant after the incident and that Thumb approached him and asked him about it. Herman had told Thumb, "There was no activity there." He admitted that it was another fireman, Merle Ironhawk, who had started the rumor. "He's the one that said things . . . pretty soon it got out. Pretty sick people. One thing you have to know is they hear what they want to hear," Thumb said. He told Jeff that he believed the company thrived on rumors and added to them.

On May 4 Jeff walked into the HR manager's office and asked him if he could type a resignation letter for him. The HR manager told him that he would do it and started laughing. "You got the last laugh," the HR manager mused. He kept laughing and Jeff joined in, secretly wondering what the HR manager would think once he found out why Jeff really came back. It felt so good to just quit.

The next day Jeff met with the operations manager, Hyllis Dauphinais, and Elvis Thumb to discuss weaving. This would be Jeff's last day of work, but they didn't know it. With Jeff's recorder turned on in his pocket, Dauphinais asked him how the machines were running. "Great," Jeff replied. "If they're running, can get everything going and then we'll be out of material," Dauphinais said. Dauphinais appeared to act dumb in front of Thumb about pick counts, something he must have been doing for a long time.

Dauphinais said that the pick problems on the machines "must have been that way all this time though. I mean we had to have shipped shit out of here on that previous contract with low picks." Jeff agreed, holding his breath and hoping Dauphinais would say more. Dauphinais continued but kept it vague, "I know Rhea made a big stink that all this material on hold downstairs here we shipped all. I know we had to have. But that stuff doesn't just change like that and all of a sudden you have low pick counts."

After the meeting Jeff told Thumb that he was quitting. Thumb was not pleased. Jeff believed Thumb had tried his hardest to get him back in the plant. He explained to Thumb that he could not work there any longer. "Come on Jeff, it will all settle down, and it will all be forgotten," Thumb pleaded. "No I can't do it," Jeff replied.

An elated feeling of liberation swept over Jeff as he drove away from Sioux Manufacturing for the last time on May 5, 2006. It was a sense of both relief and great achievement. The copies of records he had been able to smuggle out of the plant included a stack of QA weaving inspection logs that listed pick counts showing many counts lower than the required thirty-five by thirty-five yarns per inch, certifications of compliance, Kevlar purchase orders, helmet contract records, and much more. He hoped that those records, along with the taped conversations, would enhance his *qui tam* case and expose Sioux Manufacturing's fraud once and for all.

CHAPTER II

The New York Lawyer

Twenty years before he met Jeff and Tammy, Andrew J. Campanelli ran his own construction business on Long Island, New York. Born on Long Island, Campanelli was raised in a classic blue-collar, working-class family—his father was a construction worker, and his mother drove a school bus. At first he aspired to become an electrical engineer like his big brother and attended college with a major in electrical engineering. However, his life was shaped by his father's work, and he felt more comfortable in construction, so he built his own construction business and ran it for a decade.[1]

In the end, however, Campanelli realized that construction was not what he really wanted to do. Instead he had developed a keen interest in the law. He returned to college, finished his undergraduate education, and entered law school at Bridgeport School of Law in Connecticut. Campanelli worked at a law firm while attending law school and was able to finish in two and a half years. In 1992, at age thirty-one, he began his legal career on Long Island, working for a sole practitioner specializing in commercial litigation. It was a heavy-duty practice with a significant business clientele. His employer passed away suddenly in 1994, leaving Campanelli with many of the firm's clients. With those clients he started his own law practice, continuing with commercial litigation along with handling federal civil rights cases.

Situated in the predominately middle-class Long Island enclave of Mineola, about fifteen miles east of Manhattan, the office of Campanelli & Associates is located in a restored two-story early twentieth-century folk-Victorian-style style house on a frontage road at the edge of the town center. Although Campanelli entered law later than most of his peers, he possessed life experiences and street smarts that most lawyers didn't have. In 1996 he developed an interest in *qui tam* cases after reading an article in the *American Bar Association Journal* about the law. Soon afterward a whistleblower walked into his office with information about a company that was defrauding the government. Campanelli spent the following year reading everything he could about *qui tam* and False Claims Act cases. This effort led him to believe that it would be more rewarding to focus on *qui tam* cases because doing so would accomplish something for the good of everyone. By 2007 his *qui tam* practice had become a full-time venture with potential clients calling or emailing on a regular basis.

After returning from his meeting with Jeff and Tammy, Campanelli farmed out all his other cases so that he could focus on Sioux Manufacturing, an undertaking that would consume him for the next six weeks. His knowledge of Kevlar and its use in combat helmets was sketchy at best. Before traveling to Devils Lake, he had conducted a fair amount of research on both, reading articles, statistics, military specifications and testing results, and anything else he could find, mostly by using the internet searches. It was fairly clear to him that Sioux Manufacturing was underweaving the Kevlar material. He saw the danger in this practice for the troops after he found a laboratory's test results indicating that weakened weave density resulted in a sharp decrease in ballistic stopping power.

In a *qui tam* case, Campanelli believed he had to be careful what he promised a client about the success of a case. He knew that no matter how strong a client thought a case was, it could turn out to be unsuccessful in the end. As he got to know Jeff and Tammy, he felt that he was advocating not just for what was right but also for them. They had sacrificed their jobs when they could have kept their mouths shut while still receiving a paycheck and benefits. He strongly believed they were pursuing their case for the right reasons.

It is important in *qui tam* cases to develop an accurate complaint and written disclosure statement of material evidence to file in federal court. Campanelli knew that putting those documents together would require considerable planning and organization so that they would clearly state the scheme with as much specificity and substantiation as possible. In general a well-documented case that provides a high degree of substantiation lends credibility to the allegations.

"Product substitution" is the term used by the Department of Defense to categorize the type of fraud that Jeff Kenner and Tammy Elshaug were alleging about Sioux Manufacturing's conduct. Generally, it refers to contractors who deliver goods or services to the government that do not conform to contract requirements, or specifications, while seeking reimbursement based on delivery of products certified to be in conformance. In other words the contractor "substitutes" a poor-quality "product" for what was required in the government contract. The policy of the DOD is to reject any products that do not conform in all respects to contractual requirements.

Campanelli organized the complaint around the elements of the False Claims Act—existence of a claim for payment from Sioux Manufacturing's contract with UNICOR; falsity of the claim because the company had not complied with contract specifications; and presentation of the claim to the government through its prime contractor UNICOR. Sioux Manufacturing officials documented their compliance with a certificate of conformance indicating that it had complied with all contract specifications. Campanelli was well aware from his research and experience what was required in the complaint to show a violation of the False Claims Act.

"The defendants deliberately implemented a scheme to produce and thereafter sell to the DOD Kevlar shields which contained less than the minimum density of Kevlar shielding which had been deemed critical by the DOD," Campanelli wrote in his complaint, "for the specific purpose of fraudulently inducing the U.S. to pay SMC monies in amounts which would ultimately grow to equal or exceed one hundred fifty million ($150,000,000.00) dollars."

Of all the elements needed to show a false claim, determining falsity was a somewhat gray area according to court rulings. In the case of Sioux Manufacturing, Campanelli was reasonably certain that failing to comply with contractual requirements—not adhering to specifications for the weaving density of the Kevlar cloth—would render the claim as false. Although contract or specification interpretation could play a role in determining falsity in some cases, he believed that the Kevlar cloth specification did not lend itself to interpretation.

Campanelli soon received a call from Jeff, reporting in his unflappable manner that he had obtained a number of records from Sioux Manufacturing along with recorded conversations while in the plant. The first thing Campanelli did was verify that North Dakota was a one-party consent state. He had no idea what the tapes said at that point. Later, after receiving a copy of the them from Jeff and listening to them, he felt it was "pretty good stuff."

Campanelli found the taped conversations compelling enough to use as exhibits for the complaint supporting the allegations. From all the evidence he had, he felt it would have been extremely difficult to believe that Carl McKay and Hyllis Dauphinais didn't have knowledge of shorting the weave process. With physical inventories disclosing higher than normal amounts of Kevlar and notices from DuPont that the company should have lower inventories, they should have been alerted to the problem. Additionally, he believed that Rhea Crane, the QA manager, may not have known that the contract specification was not plus or minus one yarn. She had developed the quality assurance policy that called for the weave density to be thirty-five by thirty-five plus or minus one. So quality assurance thought it was perfectly proper to weave at thirty-four by thirty-four. Yet when Crane learned about the required specs, for months afterward she continued to sign certifications and allow the company to ship the Kevlar pattern sets knowing that they were underwoven.

After researching case law, Campanelli recognized two potentially key legal issues, aside from the merits of the allegations, affecting whether the case could be successfully prosecuted. The first was the sovereign immunity of the Indian tribe, which exempts Indian tribes from lawsuits generated in judicial systems that have no jurisdiction over the tribes because of their sovereign right to govern themselves. However, in many instances when tribes have business relationships with nontribal businesses, tribes may waive their immunity as a condition of doing business. Thus even though Sioux Manufacturing had waived sovereign immunity as part of its contract with UNICOR, it could claim that the waiver didn't cover charges under the False Claims Act.

Second was the question of whether an Indian tribe was a person as defined under the False Claims Act. Campanelli knew that the sovereign immunity issue would not be raised if the government intervened in the case. The question was whether the defendant would succeed with a sovereign immunity defense if the government declined to intervene. Yet rulings within the Eighth Circuit Court of Appeals, which covers North Dakota, were favorable to the case that Indian tribes and Indians would be covered as a person under *qui tam* claims. The favorable case law solidified Campanelli's decision to file the case in North Dakota.

Another reason Campanelli decided to file the complaint in North Dakota was his desire to ease the burden on Jeff and Tammy. He could have filed the case in many other federal districts outside North Dakota, but he wanted to keep it local. Before filing the case, he called Shon Hastings, the assistant U.S. attorney in Fargo, to give her advance notice of the impending filing. He gave her a brief explanation of the case, and she indicated that she would be happy to look at it. His sense was that North Dakota didn't get a lot of *qui tams*. However, now he had established a contact with the U.S. attorney's office, a good step forward.

On May 12, 2006, at a time when TBI casualties were rising among soldiers in Iraq, Campanelli submitted the complaint along with his written disclosure statement, to Drew Wrigley, U.S. attorney for North Dakota, and Alberto Gonzales, U.S. attorney general, in Washington DC. He notified them that he was commencing an action under the Federal False Claims Act against a defense contractor and its principals "who have apparently defrauded the Department of Defense."

In his notification he added, "More important than the monetary extent of the claim is the fact that, as detailed in the complaint, in defrauding the DOD, the defendants have apparently exposed all U.S. Soldiers currently fighting in Iraq and Afghanistan to a greater risk of injury or death, by deliberately and systematically 'shorting' the amount of Kevlar armor contained in their helmets, by an average of approximately 11,000 strands of Kevlar per helmet."

To make sure the Department of Justice (DOJ) promptly notified the DOD of the potentially defective helmets, Campanelli noted,

"Given the Gravity of these allegations, of which substantial evidence is enclosed, I respectfully submit that an expedited review of this claim is essential to ensure that the DOD is made aware of what has transpired and to enable the DOD to take whatever actions may be necessary to protect the U.S. soldiers who are, at present, unknowingly armed with such defective armor."

Campanelli enclosed a copy of the tape-recorded conversations between Jeff Kenner and employees of Sioux Manufacturing and a loose-leaf binder containing copies of "all material documents and information in possession of the plaintiffs."

The "Complaint, United States of America ex rel. Jeff Kenner and Tamara Elshaug, as plaintiffs" was filed in the U.S. District Court for the District of North Dakota on May 18, 2006. Along with defendant Sioux Manufacturing, co-defendants listed were the Spirit Lake Tribe, CEO Carl McKay, and operations manager Hyllis Dauphinais. Now it was the government's turn to react to Jeff's and Tammy's knowledge and concerns.

CHAPTER 12

Operation Helmet Becomes a Force

I just returned from my second tour in Iraq and was issued the old suspension style helmet system. During my last deployment, I spent time in all the worst places in the Al Anbar province of Iraq from Al Qaim to Ramadi. . . . As a Sgt., I know how the smallest thing can inspire Marines to achieve goals that would be impossible for others. Some of the Marines have talked about how the helmets sink down over your eyes when you're in the prone position or how the seat from your head will cause the old helmet to slip around your head while you're trying to run and duck for cover. Fixing this is not insignificant and can't be understated.

—Marine Corps sergeant's email to Operation Helmet

During the first half of 2006, while Jeff and Tammy were beginning their legal battle with Sioux Manufacturing over the company's shorting the Kevlar weave for PASGT helmets, and Oregon Aero was fighting the army helmet bureaucracy over its pad qualification, Doc Bob continued his battle with the Marine Corps helmet bureaucracy. By that time Operation Helmet had sent about four thousand pad kits to troops in Iraq, mostly marines in combat. The flood of requests continued unabated, while many marines complained by email to Doc Bob about the padless LWH's comfort and stability. At the same time, Operation Helmet began to garner nationwide publicity from newspaper articles and Doc Bob's appearance on CNN and other television shows and radio talk shows. Despite the credibility Operation Helmet received from media attention, the Marine Corps helmet bureaucracy, in a desperate protective mode, tried to discredit Doc Bob and Operation Helmet.

Notwithstanding the growing publicity that brought national attention to the protective benefits of a padded suspension system, the Marine Corps helmet bureaucracy continued to obstinately stand by its decision that marines in combat did not need a padded suspension system for the new LWH. This stance was not surprising, given the Marine Corps's historic culture of resistance to change.

Franz Gayl, a science and technology advisor for the Marine Corps, explained that the Marine Corps had invested a considerable amount of time and effort to develop, test, procure, and field the LWH. After the design and production of a prototype, the corps had to sell the helmet, without a padded suspension system, up the chain of command, making its case to green light the funding and establish a budget, to make the helmet a reality.

According to Gayl, once the project was funded a sort of arrogance at the higher levels of the Marine Corps became ingrained even if they hadn't actually seen the product. The helmet as sold to top leadership in the Pentagon became the reality of the helmet bureaucracy's identity. Once the project was approved, the top leaders took ownership of it. The helmet system was locked in, requirements were established, and procurements were made. Any suggestion for changes, especially from outside sources, would be resisted to protect the program and its funding line in the budget.

Thus Operation Helmet's intrusion became a direct threat to the bureaucracy's LWH program. An outsider was questioning its equipment design credibility with high-ranking officials and even its ability to provide for their troops.[1]

Former Ohio State University football star Harold "Champ" Henson was a stalwart at fullback on the 1972–74 team and an ardent supporter of the football program as an alumnus. In early 2006 Henson's son, Clayton, joined the Marines and was stationed at Camp Lejeune, North Carolina, with an expectation of deployment to Iraq. Around March 2006, with knowledge of the value of a padded helmet from his football days, Champ Henson heard about Operation Helmet and contacted Doc Bob to obtain helmet pad kits for his son and his unit.

Impressed with what Operation Helmet was doing, Henson joined the effort to raise funds for other marines so they could also have helmet pads. He told Doc Bob that he wanted to outfit every marine in Iraq. To help with fund-raising, Henson and Ohio State football coach Jim Tressel planned to stage events such as enabling fans to view a Buckeye scrimmage and contacting all alumni, including former football players, with the goal of raising significant funds for Operation Helmet. Henson also recruited friends and neighbors to help raise money. "We certainly appreciate what those young men and women are doing for our country," Tressel told the campus newspaper.[2]

When Marine Corps helmet officials learned of the fundraising effort at Ohio State, they revealed their disdain for Operation Helmet by intervening to discredit the effort. On March 14, 2006, the Marine Corps product group director for the LWH, Col. Shawn M. Reinwald, sent an email to Coach Tressel, slamming Operation Helmet. "I was sorry to see that Operation Helmet has taken advantage of you," Reinwald wrote, "and by extension, your nationally recognized football program and university, in order to advance their war profiteering." Admitting that Operation Helmet was interfering with the Marine Corps's helmet program, Colonel Reinwald continued his email diatribe: "The problem with Operation Helmet, is that the BLSS kit and their aggressive marketing campaign are interfering with the fielding and wear of the superior protection system already issued to Marines, the Light Weight Helmet.... Interference with the design and wear of the LWH by any padded system can place Marines in greater jeopardy." Then, in an effort to justify the padless LWH, he informed Tressel that the LWH was the result of extensive research and development, and its effectiveness was partly the result of the padless suspension system, which provided improved comfort and fit that reduced the "stress and fatigue of the wearer."

Contrary to the numerous complaints Operation Helmet had received about the padless LWH helmet with its lack of comfort and stability, Colonel Reinwald asserted, "Overwhelmingly positive feedback from its [the LWH's] performance [indicate that they] are meeting the demands of our Leathernecks and have eliminated ANY need for the padded BLSS kits with our deploying Marines." Despite the high rate of TBI among troops in Iraq, higher than in any previous war, Reinwald, incredibly and without citing any supporting evidence, told Tressel, "Currently, there is no definitive connection between sus-pension systems, such as in the LWH, and traumatic brain injuries." He then stated, again without providing a factual basis, "I want to reiterate that use of the BLSS with the LWH does not work. It reduces the ballistic protection of the system and does not address the inju-ries that are occurring most frequently in theater. The LWH is supe-rior to any other system available."

In addition to the misleading statements that Colonel Reinwald made to Coach Tressel, he ignored significant impact attenuation tests conducted in 2003 by a Marine Corps team at Natick compar-ing the LWH (without a padded suspension system), the MICH (with a padded suspension system), and the PASGT (without a padded sus-pension system). The results disclosed that although the LWH had an improved suspension system, "the all foam pad suspension system of the MICH is superior for impact protection."[3] The LWH's average acceleration was 158 g's, which would not pass the 150 g limit estab-lished for the ACH and the MICH, while the MICH averaged 113 g's, a 72 percent impact attenuation over the LWH.

Regrettably, Reinwald's email had its intended effect, causing Tressel to cancel his fund-raising plans. However, Champ Henson, angry and appalled at what the Marine Corps had done, made his concerns known to Marine Corps officials in Washington DC. They apologized and expressed disgust at the email and the inference that Operation Helmet was a war-profiteering organization. Doc Bob's response was to not back down or stop sending marines the helmet pad kits they sorely needed, even more so now. Reinwald, in turn, sent an email to Doc Bob, arrogantly telling him, "You are not a shipmate" and again accusing Operation Helmet of "abetting war profiteers."

One day in the spring of 2006, Doc Bob was sitting on his back porch and thinking about a round of golf when he received an unlikely phone call. "Hi, I'm Cher," came the voice over the phone. "I heard you needed some help." Initially responding with some skepticism, Doc Bob finally determined it really was the famous singer and actress. He asked her how she had heard about Operation Helmet. Cher explained that her sister had seen an article in a California newspa-per written by a Marine Corps gunnery sergeant's family, referring to the good work of Operation Helmet. Her sister had torn out the article and sent it to her. She had written across it, "Do something stupid." Feeling she needed to support the organization, Cher looked up its website and phone number. She told Doc Bob she wanted to send him some money and asked how much he needed. After some discussion Cher sent Doc Bob about $35,000. A couple of weeks later, fearing that her contribution was not enough, she called Doc Bob again and then sent him $100,000. Her donations made a healthy dent in his waiting list.

Within a day or two of Cher's phone call, Doc Bob's son, Mark Meaders, who lives in the southern part of Pennsylvania, near its border with Maryland, met with his congressman, Representative Curt Weldon (R-PA), about Operation Helmet and the benefits of the padded helmet suspension system. Mark had fifteen minutes on Weldon's schedule to talk about the subject. With two helmets in hand, he gave Weldon a five-minute talk explaining that one helmet, without the padded suspension system, takes a 300 g hit that will transfer 157 g's through to the brain, while the other helmet, with a padded suspension system, takes the same impact but will transfer only 78 g's through to the brain, the difference between a serious TBI and no injury.

Seemingly stunned by this revelation, Weldon, who at the time was the chairman of the House Armed Services Committee's Tactical Air and Land Forces Subcommittee, immediately called a committee staffer into the room. They went through the same presentation again. Weldon then asked for a hearing with the committee. The hearing would center on the helmet pad issue and feature the testimony of Doc Bob, representing Operation Helmet. It was a watershed moment for his organization.[4]

The hearing was set for June 15, 2006. When Cher learned of the hearing and that Doc Bob would testify, she called him and asked if he had ever done anything like that before. He said no, and she then offered to accompany him to the hearing. Doc Bob immediately accepted. He knew Cher's presence would add to the excitement and, more importantly, bring attention to the troops and the helmets.

Ironically, a month before the hearing, in May 2006, the Marine

Corp's Center for Lessons Learned published a report on management of TBI in the Iraq theater of operations. Among its recommendations to Marine Corps leadership was the following: "The design of helmets, helmet webbing and chinstraps, combat arms ear protection (CAEP), restraints in vehicles commonly used in convoys and exposed to IEDS, and other non medical protective equipment should be evaluated for potential design modifications to ameliorate or prevent TBI from blast injury."[5] It appeared that the leadership had ignored this recommendation.

CHAPTER 13

Feds Move In

D avid Peterson, the chief of the Civil Division, had more experience with civil cases than any other assistant U.S. attorney in North Dakota's U.S. attorney's office. He was working in his Bismarck office on a spring day in mid-May 2006 when he received a call from Shon Hastings, the assistant U.S. attorney at the branch office in Fargo. She had received the complaint and disclosure statement from Andrew Campanelli and wanted Peterson to read it. *Qui tam* cases were rarely seen in that office. After Peterson had reviewed the case information, it was clear to him that Campanelli had done his homework.

Peterson was an Army veteran who had worn the old "steel pot" combat helmet in the late 1950s and early 1960s. As he read the complaint, he quickly became concerned about the safety of the PASGT helmets if they contained less Kevlar than required. Also, he had friends and relatives who had kids in Iraq or Afghanistan wearing those helmets. He didn't like the idea that the Army might not be getting what it had contracted and paid for. After reviewing the case information, he felt there were two issues that needed to be addressed—does the amount of Kevlar comply with what the contract called for, and would the helmets pass ballistic testing?

When he first started work in the U.S. attorney's office in 1972, Peterson opened the agency's office in Bismarck with just a secretary and him. In 1976 he left federal service and entered private practice in Bismarck. Trial work was his first love. It was what he enjoyed most and was able to practice while representing plaintiffs in personal injury, medical malpractice, and product liability cases during his years in private practice. However, in the 1990s such cases were increasingly being settled out of court or through arbitration, with few going to trial. After nineteen years in private practice, Peterson wanted to get back into the U.S. attorney's office so he could try cases. An opening occurred in 1995, and he was rehired. When Peterson returned to the U.S. attorney's office, staffing had increased to six attorneys in the Bismarck office alone, along with the attorneys located in the Fargo office.

In addition to civil cases, Peterson handled many criminal matters that occurred on Indian reservations in North Dakota for the U.S. attorney's office—murder, kidnapping, rape, child and sex abuse, the whole range of major crimes. He tried three first-degree murder cases. One of those cases resulted in eleven people going to prison for everything from first-degree murder to aiding and abetting perjury before the grand jury. Before the Sioux Manufacturing case was filed, he had never handled a *qui tam* case. He had looked at a few such cases, but they were inconsequential nonstarters. His first step was to arrange a meeting with Campanelli and the relators at the Fargo office, for May 17 and 18.[1]

The U.S. attorney's office in Fargo is located on the second floor of the older of two connected but architecturally diverse, buildings—the oldest federal building was built in 1931, and the newer one, a modern edifice, where the courtrooms and judge's chambers are located, was added in 1998. The complex is named the Eugene Burdick Courthouse and Federal Building after the man who served the State of North Dakota as a U.S. representative and senator for thirty-four years.

On May 17, 2006, nervous and unsure of what to expect, Jeff Kenner and Tammy Elshaug walked into the U.S. attorney's office, where they met with Campanelli, who had flown in from New York the night before. Tammy recalled thinking, "What are we doing, what are we getting into?" It was an intimidating environment for two small farming-town citizens who had never interacted with federal authority. They were ushered into a conference room where Peterson and Shon Hastings were waiting to ask them detailed questions about the case.[2]

As soon as they were settled, Peterson explained to Jeff and Tammy that he wanted to interview each of them separately and Jeff would be the first to be interviewed. After Tammy left the room, Peterson walked Jeff through the facts and evidence disclosed in the complaint. The questioning was very detailed and elicited important information that Peterson felt he could use in the investigation. It was important for him to know not only the facts and evidence of the case but to understand all the technical aspects of the weaving and thread-counting process and the quality assurance inspection procedures. He also wanted to take measure of both relators. Jeff masterfully provided information to him in a calm and rational manner.

Peterson and Hastings took copious notes of Jeff's information with a view toward planning how to proceed on the case and determine which records they needed to obtain and whom to interview. Jeff's interview took all day, which meant that Tammy would be interviewed the next day. Staying overnight in town was not a problem. Her sister lived in Fargo, and she could stay at her home. The process was repeated the next day with Tammy. Again, detailed questions were asked of her, and she provided the information based on her knowledge of the case.

During that first meeting, Peterson queried both Jeff and Tammy about the alleged affair. At one point when they were not in the room, Peterson turned to Campanelli and asked him, "Do you know about these allegations of an affair? We've got to know what the truth is." "Look, I haven't known Jeff or Tammy that long," Campanelli explained, "but for the length of time I have known them, to me, they seem to be honest and forthright." Peterson pointedly asked, "Do you believe them?" "Yes, 100 percent," Campanelli confidently replied. "If I had any doubt, I would tell you, attorney to attorney, I have doubt. I have no doubt they are telling the truth." Jeff and Tammy were then brought into the room and asked about the affair. Both were emotional about it, but said no, it was a lie that Sioux Manufacturing managers had just fabricated.

Peterson did not believe the allegations of an affair between Jeff and Tammy. However, it concerned him because it could be a big issue that Sioux Manufacturing would bring up later. In order to be sure about the issue, he had investigators do follow-up on the information they had given him. In the many interviews the investigators eventually conducted, he found nothing that substantiated the affair allegations. Based on the investigative results, Peterson concluded that the affair did not happen and the issue was moot.

On May 18, while attending the first relator meeting in Fargo, Campanelli formally filed the complaint, titled "United States of America ex rel. Jeff Kenner and Tamra Elshaug, v. The Spirit Lake Tribe, Sioux Manufacturing Corporation, Carl R. McKay and Hyllis Dauphinais, with the United States District Court, District of North Dakota, Northeastern Division, Fargo, North Dakota, under seal [without public disclosure], as is required pursuant to the qui tam provisions of the federal False Claims Act." The complaint was not served on the defendants during the seal period. The *qui tam* law requires that the complaint be filed under seal for an initial period of sixty days to allow the DOJ time to investigate the allegations to determine if it should intervene. Also, the seal period may allow the DOJ or the U.S. attorney in the district where the case was filed to conduct a criminal investigation, if warranted, to obtain hard evidence through the use of available investigative tools. Normally, the DOJ will request that the court extend the seal period so it can conclude its investigation. Extensions of the seal sometimes can lengthen the investigation of the case for several years.[3]

Maintaining the integrity of the seal is important in any *qui tam* action. A violation of the seal that results in a premature disclosure to the defendants could thwart the government's investigation, preventing it from obtaining important evidence. Premature disclosure could also lead to a dismissal of the case. Once the complaint is filed, the DOJ or the U.S. attorney normally assigns the case to an investigative agency with jurisdiction over the allegations to conduct an investigation leading to a determination whether the DOJ will intervene and prosecute the case.

While a case is under seal, the DOJ has a number of options in making its decision to resolve the case. It can elect to join in the lawsuit, decline to join, move to dismiss the action, or attempt to settle prior to a prosecution. If the DOJ elects to join, it controls the action and has the primary responsibility for prosecuting the case. If it declines to join, the DOJ will pull out of the investigation, but the relator is entitled to investigate and prosecute the case at his or her expense.

Jeff, Tammy, and Campanelli left the meeting in Fargo feeling good about the interview. Peterson and Hastings seemed very positive toward the case and had said that it was the most impressive false claims action they had seen. To Jeff and Tammy, it was exciting and scary at the same time. The relators had impressed Peterson as "being honest in their concern about what they thought they knew was wrong at Sioux Manufacturing." He had worked with a lot of witnesses and had a pretty good instinct for recognizing the good ones and the bad ones. He believed Jeff and Tammy were sincere in their motivation and would make good witnesses.[4]

Peterson returned to his office and called the DOJ to coordinate the case with an investigative agency having jurisdictional responsibility. Since the prime contractor in the case was UNICOR, a component of the DOJ, the department put him in touch with the DOJ Office of Inspector General (DOJ-OIG). The DOJ-OIG had been established under the Inspector General Act of 1978 to conduct independent investigations, audits, inspections, and special reviews of DOJ personnel and programs in order to detect and deter waste, fraud, abuse, and misconduct associated with DOJ operations. Within the DOJ-OIG the Investigations Division investigates alleged fraud involving DOJ contractors. Thus it was the component assigned to this investigation.

The DOJ-OIG, in turn, organized a task force that included the DOD investigative agencies that also had jurisdiction—the Defense Criminal Investigative Service (DCIS), the investigative arm of the DOD inspector general, the Army Criminal Investigations Division (CID), and Air Force Office of Special Investigations (OSI). Special agent Ken Dieffenbach of the DOJ-OIG was assigned as lead agent on the case. Once the investigative team was set, Peterson decided the first investigative task would be to execute a search warrant on Sioux Manufacturing to obtain records and other evidence needed to prove the case.

The decision to conduct a search of Sioux Manufacturing was not made lightly. It required considerable manpower, coordination, and planning. It would also have the effect of shutting the business down for at least a day. Peterson had the option of the DOJ-OIG simply issuing an inspector general subpoena for the records. However, given the concern he had that the evidence could be destroyed, hidden, or altered if the company was given time to respond to a subpoena, he felt it necessary to go in and get the records as they existed without advance notice. Additionally, the search would provide the investigators the opportunity to conduct onsite employee interviews and offsite interviews of former employees.

It was important to Peterson not to tip the company's hand on the search and that meant not revealing it even to Jeff, Tammy, or Campanelli. At the same time, it was also important to garner detailed information about the physical layout of Sioux Manufacturing to assist their planning of the raid. On June 2, 2006, Shon Hastings emailed Campanelli with numerous questions for Jeff and Tammy to answer. The request did not mention that the information was being obtained to execute a search warrant. The questions centered on the location of quality assurance records, accounting documents, weaving maintenance logs, along with the identification of former Sioux Manufacturing employees, security features at the facility, and hours of operation.

Hastings also asked that the relators draw and label a diagram or map of the physical layout of the plant. A meeting was scheduled for June 14 to include representatives from all the agencies that were to participate in the search, along with Jeff, Tammy, and Campanelli. By June 6 Peterson and Hastings had received written answers from Jeff and Tammy to their questions about the plant. Not long afterward they received the detailed diagram of Sioux Manufacturing. The map of the plant identified where everything was located including every office, storage rooms where records were located, and the location of each computer. They identified at least one hundred items inside the plant. By the time of the meeting, all the investigators knew exactly where everything was located.[5]

At the June 14 meeting in the U.S. attorney's office, Campanelli recalled that the room "was very crowded." Jeff and Tammy again walked the investigators through every detail of Sioux Manufacturing. Although it was not mentioned, Campanelli sensed from the questioning that a search was going to be conducted. Concerned about possible retribution if a search were to take place, he took Peterson aside after the meeting and requested that if a raid was conducted,

Peterson would call him and give him a heads-up that "something [was] going to happen," so he could warn Jeff and Tammy. Peterson assured Campanelli he would do so if the investigators were going to search the Sioux Manufacturing plant.[6]

On June 20, the day before the raid, all the federal agents involved in the search, along with Peterson, met in Grand Forks to coordinate the facility search. Ken Dieffenbach from the DOJ-OIG was lead agent on the search. Peterson was impressed with his efforts to put the whole search scenario together. About thirty-seven agents from all the task-force agencies attended the meeting. Dieffenbach had the diagram of the facility that Jeff and Tammy had drawn. He superimposed everyone's responsibilities at the plant on the diagram. Peterson felt that Dieffenbach had the search operation well organized, and they were ready to go into action the next morning with the search warrant signed by Federal Magistrate Karen Klein.

Around 5:00 p.m. that same day, Campanelli received a call from Peterson, who told him, "Something is going to happen tomorrow. I just wanted you to know." That cryptic statement was all he said. Campanelli thanked him. He knew what it meant. He immediately called Jeff and alerted him that Sioux Manufacturing was going to be raided the next day.[7]

Early the next morning, at least thirty-seven federal agents from the DOJ-OIG, DOD investigative agencies, and the FBI gathered in Grand Forks, prepared to commence the raid on Sioux Manufacturing. Devils Lake was almost a two-hour drive from Grand Forks, and the investigators wanted to arrive at Sioux Manufacturing by 9:00 a.m., soon after the company's first shift of the day began. Peterson recalled the first glitch of the day when one agent from the air force base in Minot, who had not attended the coordination meeting the previous day in Grand Forks, showed up carrying a shotgun, something no one else had because it was not needed.

When the investigators arrived at Sioux Manufacturing at 9:00 a.m., they first barricaded a small road just outside the plant. Then they immediately served the search warrant on the startled CEO, Carl McKay. He was told that the business would be shut down during the search, and the employees would be herded into the cafeteria to be available for interviews. The agents, divided into three teams, entered the plant. The first team went after the records. Thanks to Jeff's and Tammy's help, they knew in advance where the pertinent records were located. Once they found the records, they were loaded into boxes and carted to one of two rented trucks. The second team, composed of computer experts, downloaded hard drives from targeted computers. The third team conducted interviews of the employees. Later in the day and the evening, this team would spread out around the reservation to locate and interview as many former employees they could find, mostly at their homes.[8]

The search produced thousands of pages of records and much of the company's computer hard-drive data. Peterson considered the search well organized and successful. He believed the investigators had obtained all the records they were looking for to prove their case. Campanelli recalled Shon Hastings telling him that they had found exactly what they thought they would find.[9]

Soon after the search ended, the company HR manager called Jeff to tell him that the plant had been raided. Jeff wanted to tell him how the raid originated, but he couldn't talk about the filed lawsuit. The HR manager told Jeff that the federal agents came in with guns, rounded everybody up, put them in a front room, and selected people for questioning. He then revealed that McKay told management at a post-search meeting that it had been Jeff's doing and that he was going to pay for the company's lost time and production. All Jeff could do was laugh.[10]

Tammy was shopping in Devils Lake on the day of the raid and had run into a woman she knew who was a production worker at Sioux Manufacturing. The woman asked her if she had heard that the plant was raided. Tammy pretended she knew nothing about it, but she felt anxious. She believed that Sioux Manufacturing wouldn't know who was behind the raid since the complaint was filed under seal. However, she feared retaliation if the company found out. Tammy was afraid that people intent on retaliation would be outside her home when she went to work at the post office in the middle of the night.[11]

The next morning agents drove two rental trucks loaded with thousands of pages of records seized in the raid back to Fargo. The records were placed in the basement of the federal building, and investigators started poring over them.

"Sioux Manufacturing Plant Raided by Federal Agents" headlined the morning local newspaper, the *Devils Lake Journal*, on June 22, 2006. "Federal agents raided an American Indian–owned armor plant based in Fort Totten on Wednesday, confiscating boxes of company records under a search warrant involving questionable business practices," the article read. "Assistant U.S. Attorney, David Peterson said Sioux Manufacturing Corp. was under investigation by the Justice Department and the U.S. Attorney's office for alleged contract irregularities."

"About 25 to 30 agents came in with their shotguns out and locked down the plant," Sioux Manufacturing CEO, Carl McKay was quoted as saying. "It just scared the dickens out of our employees. . . . We have nothing to hide. . . . No one was arrested or read their rights or anything like that." McKay was also quoted as blaming the raid on "some disgruntled former workers."

Peterson would later recall that the allegation that the agents came into the plant with their shotguns was ridiculous. The operation was tightly organized. Searches of companies, such as Sioux Manufacturing, that are being investigated for an alleged fraud against the government, are normally carried out in a methodical manner with the goal of obtaining evidence such as records that could be used to prove the case in court. Federal agents, like any law enforcement personnel, carry handguns. Unlike in a drug raid or similar action that would likely present danger to the agents, the agents' weapons are holstered in a corporate environment.[12]

Soon after the raid, a Sioux Manufacturing employee walked into Goldings, a Devils Lake store where Jeff worked. Seeing Jeff, the man asked him what he knew about the raid. Jeff played dumb and told him he didn't know anything about it. "Well, I heard that if you show up on the reservation, you'll be skinned alive," the man warned. Jeff shrugged off the remark and just laughed, but he knew that the threat originated from someone in management. It was the first time that he felt he had better not go out to the reservation. The gravity of his whistleblowing was beginning to sink in.[13]

More than forty thousand documents and over thirty computer

hard drives were seized during the raid. This material was placed in the basement of the federal building in Fargo, and investigators quickly cataloged and inventoried it. Assistant U.S. Attorney Shon Hastings told Campanelli that they worked faster than on any other case she had seen. Along with the analysis of the documents and the hard drives, investigators interviewed approximately seventy-four witnesses and consulted with industry and military textile and armor experts. According to their findings, Peterson believed, the investigation largely substantiated Jeff's and Tammy's allegations that Sioux Manufacturing "did not always comply with the thirty-five-by-thirty-five thread count standard in its manufacture of Kevlar helmet cloth." Additionally, "witnesses disclosed that some inspectors improperly counted and/or rounded their thread counts and that some employees were provided with incorrect training related to conducting thread counts."[14]

Despite evidence-based findings, questions remained in Peterson's mind. What was the safety impact to the PASGT helmet? Were any injuries or deaths caused by the failure of the PASGT helmet to provide ballistic protection? Also, what actions was the Army going to take as a result of the problem? With potentially millions of substandard PASGT helmets in the field, he believed that the Army should have immediately addressed the issue.

Even though the U.S. attorney's office in North Dakota was leading the investigation, the DOJ had final authority on whether there would be government intervention or declination of the *qui tam* case. An intervention would mean that the DOJ would proceed to prosecute the case, while a declination would mean that the DOJ would pull out of the case, leading to a dismissal or allowing the relators to proceed in court. To get a good feel for the case and the relators, the DOJ sent its senior attorney for the Commercial Litigation Branch, Dennis Phillips, to Fargo, on September 12, 2006, to interview the relators and review the case.

Jeff and Tammy, along with Campanelli, walked into the meeting with Phillips at the U.S. attorney's office in Fargo. The room was filled with other interested parties including Peterson, Hastings, and agents from the investigative team. Campanelli realized that Phillips's presence indicated that the DOJ was serious about the case. He believed that if the case went to court, Phillips would be the lead trial attorney for the DOJ. He was there to size up the evidence and to determine the credibility of the relators. Phillips had considerable experience in handling large cases, especially those involving defense contractors. Campanelli was thrilled that Phillips would be involved, given his reputation for successfully prosecuting big cases.

Jeff and Tammy were interviewed separately. Phillips interviewed Jeff first. "He was tough, all business," Jeff recalled. "He really grilled me." He couldn't get a read on whether Phillips had a positive feeling about the case. Tammy's interview began in the afternoon and then continued the next morning. She was asked how they came up with the dollar amount ($150 million stated in the complaint). She didn't know and referred Phillips to Campanelli to answer the question. Tammy got the feeling that Phillips didn't want Campanelli to be the person answering the question. The interview made her feel uncomfortable, but she wasn't stumped by any of the factual questions. During a break in the interview, Phillips commented to Peterson that even if the ballistics passed a test, the DOD didn't get what it had paid for. This could be a basis for a false claim to the government just by itself.[15]

Peterson felt that Phillips knew what he was doing. He showed his knowledge during the interviews, going through the complaint and pointing out the elements he believed were important and what he thought was less important. He appeared to have a firm grasp of the complicated *qui tam* law and what was required for a successful prosecution. Phillips told Peterson the case needed considerable additional investigation. He thought the relators were legitimate, concerned citizens who had a concern about the product.

Unfortunately, the meeting and interview with Phillips was the last time Peterson ever dealt with him. Shortly afterward, he learned that Phillips had left the DOJ to accept an administrative law judge position in Florida. The contact Peterson was given after Phillips left was a young woman attorney who had almost no experience and was fresh out of law school. She was at the DOJ for only a short time. Eventually, another senior attorney became his point of contact at the DOJ.[16]

CHAPTER 14

Breakthrough for the Troops

June 15, 2006, was a typical humid, early-summer Washington DC morning when Doc Bob, his son, Mark, and Cher, strode into the House of Representative's Rayburn Building, where the Tactical Air and Land Forces Subcommittee was set to hold hearings on the status of combat helmets along with body and vehicle armor in Iraq and Afghanistan. It was a watershed moment for Doc Bob and Operation Helmet. He was scheduled to speak, and Cher was there for support. They made an attention-grabbing pair.

In the anteroom before the start of the hearings, the Army, Navy and Air Force speakers warmly greeted Doc Bob, but the Marine Corps speaker, Maj. Gen. William Catto, commanding general, Marine Corps Systems Command, remained aloof and generally ignored him. Cher circulated about the room, commanding most of the attention.

As they entered a packed hearing room, Cher created a big stir—the cameras were trained on her, suddenly thrusting Operation Helmet into the limelight. The hearing drew the largest crowd ever to be in the room. As one committee member said, "If we were having a hearing on this issue without the presence of a celebrity, we would not have the number of photographers that join[ed] us." Chairman Weldon dragged Cher around by the hand, proudly showing her off to his colleagues in the room. It was a politician's dream press event.[1]

The hearing was divided into two panels of speakers. The first panel consisted only of Doc Bob. The second panel included a number of senior military officials. Roger Smith, deputy assistant secretary of the Navy for littoral and mine warfare, and Maj. Gen. Gary McCoy (USAF), director of logistics readiness, Office of the Deputy Chief of Staff for Logistics, Installations and Mission Support, were the main witnesses, while their staff filled the chairs behind them. Doc Bob testified first. He took his seat at the speaker's table with his name prominently displayed in front of him. Cher sat directly behind him.

Chairman Weldon brought the hearing to order and began with an opening statement disclosing that available test data of helmet suspension systems, although limited, indicated "significant differences" between padded and sling suspension systems with "the padded suspension system providing approximately twice the protection against blasts."[2]

Weldon then referred to the Marine Corps's infamous point paper that tried to make the case for the Corps's position of not authorizing the padded suspension system for its LWH. He asked if they were "misinformed." "Apparently we have thousands of our military personnel who believe the helmet they are being issued does not provide them satisfactory protection," Weldon told those at the hearing, referring to Operation Helmet. "When we asked the Marine Corps why some Marines were expressing dissatisfaction with their helmets, the official position was that the Marine Corps helmet provided the required protection." Weldon then chastised Marine Corps senior officials for claiming that "there was an inappropriate relationship between Operation Helmet and the primary provider of the padded system for combat helmets. In fact, a senior Marine Corps official accused Operation Helmet of abetting war profiteering," he added.

Weldon wondered why thousands of military personnel in combat had expressed a need for a padded suspension system for their helmets: "How can so many war fighters be wrong? The Army and Special Operations Command use the padded system being requested. The Marine Corps' own testing indicates that their helmet provides about half the blast impact protection of the Army's [inaudible] helmet. The Marine Corps says its helmet meets the Marine Corps requirement," Weldon continued, "but if it only provides half the blast impact protection of the Army helmet, we need to understand why this is acceptable to Marine Corps leadership and why it insists on using the sling suspension system."

Weldon concluded his opening statement: "It makes no sense of [*sic*] me that our military personnel have to rely on a charitable organization to get the equipment they seem to think they need because the service is not providing it for them." Then he asked a pointed but logical question: "If our troops are asking in great numbers, as appears to be the case, and I understand there have been 8,000 inserts sent over from this foundation which are being used by our troops, then why don't we issue this to them?"

Representative Solomon P. Ortiz (D-TX), the ranking member present at the hearing, provided his opening statement: "Three years into the war in Iraq, it is a shock to no one on this committee that the different services tend to go their separate ways on buying helicopters, airplanes or other expensive equipment. In some cases there are good reasons for the services to have different equipment. But, in the case of a helmet for ground combat, any argument that different gear is required for the Army and Marines strains the bounds of credibility. It is wasteful and it means that one or the other, either the Army or the Marines, are receiving inferior force protection equipment."

Following the opening statements by committee members, Doc Bob presented his testimony. Hearing protocol requires a speaker to read his or her testimony within a five-minute period. This oral testimony is normally a summary of a longer prepared statement submitted in advance to the committee. Witness testimony is followed with questioning by each of the committee members.

Speaking with confidence, Doc Bob told the committee members how he had become aware of helmet pads and started his organization to provide an upgraded pad kit to the troops. He noted that he had provided over eight thousand kits to date. He described his research of the pads and how it had convinced him, as a doctor, of the pads' importance for the troops in combat. He praised the help he had received from families, organizations, and others who raised money or donated to his organization to provide the troops with helmet pads.

In response to questions by the committee members, Doc Bob spoke of the many emails he had received from troops, especially marines, complaining about the wearability of their helmets, both the old PASGT and new LWH, and their desire to upgrade their helmets with pads. He gave examples of the many anecdotes the troops had shared in their emails, including one from a young marine corporal who was in the Camp Lejeune Naval Hospital recovering from shrapnel wounds to his neck and shoulders. "When the rocket hit the wall behind me and exploded, it shredded my helmet completely and filled my shoulders and neck with shrapnel," he wrote to Doc Bob. "They're taking care of that. But the good news is, it didn't hurt my head at all. It just destroyed the helmet and the pads probably saved my life."

"I don't want to put myself across as a brain injury specialist, because I'm not," Doc Bob recalled of his experience with combat injuries while serving in Vietnam. "But on a practical level, we saw them [soldiers with brain injuries] and we treated them and we saw the devastated effects of them. And seeing one is enough to last you for a lifetime. And if there is a way that subsequently we can do something to prevent it in the first place, then, good."

Doc Bob continued:

So, anecdotally, we say, yeah, it works, and it's a hard thing to prove, I guess. You know, if you stood up ten guys in a row and hit them, each one a little bit harder, until one of them died, then you could say this is the level of protection that we need. But, so, we have to go on information from the Brain Injury Association, who is here with us today, from the National Football League, that mount G-force meters in their helmets and show just how much people get whacked in the head, and soccer players that mount them in their headbands. So, there is pretty good information out there about what levels of impact forces will result in headache, loss of consciousness, and ultimately fatal injuries.

Addressing an issue used by the Marine Corps against Doc Bob, Weldon questioned Doc Bob: "As I mentioned in the opening statement, and I want to repeat this for my colleagues and for the public, all the money you raise for this foundation, you put out the backdoor to buy the inserts. Isn't that correct?" "Yes, sir," Doc Bob responded, "We don't take a dime in salary or expenses that way."

Clearly angered at the Marine Corps, Weldon told the hearing that it was "pretty outrageous" to hear that the Marine Corps would try to tie Operation Helmet to a vendor. "You're probably as disgusted as I am," he said, speaking to Doc Bob, "that someone would even suggest that." Doc Bob responded, "Yes, we have purchased from several vendors, and we winnowed it down to the one, basically, that I would want on my head if I were frontline combat troops, and I would want on my grandson's head." Doc Bob stressed that he was shocked the military hadn't recognized the need for a padded suspension system over the years.

When Representative James Gibbons (R-NV) took his turn, he stated, "I am stunned, stunned that our military doesn't have the foresight to see the need for an insert like this and has not seen the need for an insert like this over the years. We have looked at or watched the evolution of the National Football League and their helmet systems over the years. They have been out front, leading the way. You would think that some of the knowledge base, the institutional knowledge base of helmet design and protection of the head would have been transferred from those professional institutions to our military."

After several questions relating to cost and the use of the National Institute for the Blind to procure pads for the Army, Gibbons made his concluding remarks:

> The word of mouth that's out there in our military among the troops is very effective in terms of awareness today of the improvement in their helmet by what you have done, or what you have provided them, and it stuns me again that the industry that's responsible, the Department of Defense, which is responsible for that industry and the soldiers, haven't taken the opportunity to improve the helmets they have out there. To me, I think there is a clear choice here. We either protect the lives of our soldiers or we don't send them. And if we have the ability to protect them, we ought to do everything possible to do that.

Chairman Weldon then had the last word for Doc Bob by praising his testimony:

> Well, Dr. Meaders, you did well in your first appearance before Congress. . . . You wowed them, I guess, is the word I would say. No negatives, all positives, and some questions, legitimate ones, raised. . . . And

we have the same concern, because we want to make sure that what we're doing, as you do, is in the end right and proper. But, you know, no one can question your patriotism and the service of you and your family to the country. You're an amazing group of people. And with that, I'd like to ask you, your wife, your son and his wife to stand up, along with Cher, so that we can give you all a special round of appreciation as you leave the witness table.

With that everyone in the hearing room stood and applauded Doc Bob and Cher. It was a singular moment of respect, especially for Doc Bob's efforts.

Following Doc Bob the second witness panel was convened with military officers, decked out in their dress uniforms full of bright gold stars, medals, ribbons, and other accoutrements of military status. Behind them sat a row of military aides, available to provide them any information they needed. These aides are typically present during military testimony and are known as part of "the flotilla" by congressional staffers.

The Army's Maj. Gen. Stephen Speakes, director of Army force development, testified by lauding the Army's new ACH, "which is drawn from the Special Operations Command. It has a very different suspension system in it," referring to new improvements. "The suspension in it is a pad-based system, which we believe is much more effective at resisting concussion, which is another area that we saw a problem."

In contrast to the Army's helmet, the Marine Corps's LWH had no padded suspension system. In his testimony Maj. Gen. William Catto, outgoing senior acquisition officer for the Marine Corps, praised the Corps's new, padless LWH and its ballistic protection capabilities while avoiding the issue of nonballistic and blast-impact protection. In vague military bureaucratic language, common with military presenters, he claimed, "We're doing absolutely everything we can to ensure the safety of our marines by providing them with the world's best and most effective force protection solutions."

In response to Catto's testimony, Weldon lectured him on the fact that nonballistic impact tests had demonstrated that a helmet with a padded suspension system provided twice the protection of a helmet without the padded system. He asked Catto why he wouldn't want that protection for his marines? Trying to articulate a justification, Catto responded that they did purchase padded helmets for the Marines' "reconnaissance folks and for [Marine Corps] people that are in the Air Naval Gunfighter and Liaison Crews, because they're guys that do either direct operations that need that kind of capability . . . and parachute ops, so that when they fall or have the kind of blunt-force trauma that they may have through parachute ops, that's a very good helmet." He did not consider the same protection for the regular marines in combat and also in "direct operations," as if these troops were not worth the cost of a padded suspension system.

Catto went on to identify three areas he needed to consider for improving the helmet: "There is ballistic protection, there is crash protection, and then there is a protection from the results of a concussion or percussion. We don't know about the blast piece, the concussion, and we need research for that. But all I've heard for the ACH from everyone is that it's better against the crash and it's more comfortable. I have to have the rigorous data that tells me what's the best solution for blast, for concussion, you know, that piece, for crash and for ballistic protection. And that's why I say, I need to help on the research to get to that." However, later in his questioning Catto would admit the Marine Corps did not have the funding to conduct blast research.

In response Weldon contended, "The Army is already using the helmet. They've obviously addressed all three areas that you just said were the important elements for the Marine Corps. The marines have voted with their mouths and are telling Dr. Meaders [to] send us the helmets." Although both the Army and the Marine Corps had conducted blunt impact tests and Doc Bob had sent General Catto blunt-impact test results for pads in November 2005 indicating that padded suspension systems provided much better head protection over padless systems, Catto continued to claim that he did not have the necessary information to make a decision. Displaying perhaps an institutional ignorance of the issue, Catto continued to assert, "I need the facts. I don't have the facts. I don't have the fact on what is happening in the combat trauma registry for what's really happening to our guys."

Weldon was not through with Catto or the issue. He replied to Catto, "General, I don't want to beat a horse to death, but we have a copy up here of Oregon Aero's test results [the same results that Doc Bob had sent Catto]. Has the Marine Corps asked them for their test results?" "Sir, yes we have. I have not seen those results. We asked for them in September," Catto replied. "Here they are," Weldon retorted. "And they're comparative results. And what we're told here is these are basically the same evaluations with the Marine Corps lightweight helmet. And the results are pretty much what we've heard the Army state to be the case. You know, it is hard for us to be here and to understand all the dynamics in question here. What we do know is very simple. We do know we want the best protection."

Weldon then issued a direct challenge to Catto to "make my day" by agreeing to provide helmet pads to all marines in combat, not just the reconnaissance units, while any studies were being done. "We will give you the money," Weldon told Catto. "Let's put Dr. Meaders out of business. Let's buy these inserts and make them available for the troops today. We're with you. Democrats, Republicans, you've got the money. Buy them. Let the marines use them as they're doing. And then if there is a study that shows we should improve it another way, then fine. To me, if it makes the soldiers who are using these inserts feel comfortable, then we ought to do that. You do the study, but let's do that now. And we're not going to take this out of the Marine Corps budget."

Still resistant, Catto replied:

Congressman, you've always been a great supporter and your heart is pure and I love that. I will not go ahead and authorize the use of those pads unilaterally until I have the data that says what the right decision is. Now, having said that, I'm not going to tell them they cannot use it, but the issue we've talked about here has been primarily comfort or a better protection against crash. I've got to have the data to make sure that we make the right decision before we as a service move one way or the other. And I'm not trying to be a roadblock here.

Weldon voiced his displeasure:

Well, respectfully, General, I would say the Army is looking at the same three areas that you're looking at. They're not just looking at ballistics, they're looking at crash, they're looking at everything. And they've made a decision, and I understand that recon has a different function and you've made that decision for them but not for the others. But I can tell you; the decision in my opinion has already been made. It's been made by the marine on the ground. And this wasn't something forced on them. There was no marketing team over in the theater saying hey, buy this. This was a group of Americans who said we'll raise the money to give you, if you want it, and 6,000 marines now have that. So, I'm going to tell you what I'm going to do. We're either going to raise the money privately and I'm going to get behind Dr. Meaders and do it, and if we do that, it's going to embarrass the service. We shouldn't have to do it. Or we're going to force it through an appropriations process that this Congress has the ability to do.

Weldon concluded the hearing, seemingly frustrated at Catto's resistance to supplying a padded suspension system for all marines. "It just defies logic," he said, "for us not to know why this is not being done. And to have to go out and continue to beg and borrow and raise money from the public is not the answer. It's not what the American people want to hear."

General Catto did not rise to the rank of major general by being ignorant, but he clearly was myopic about this issue. He obviously was a loyal marine. Yet his stubborn resistance to overwhelming evidence that a helmet with a padded suspension system had important protective benefits for the marine in combat suggested that his subordinates or the helmet bureaucracy had possibly kept him in the dark or perhaps he had never examined the blunt-impact test results that Doc Bob had sent him. It appeared that he was defending the decisions he was aware of, or what he was told, by the helmet bureaucracy and its point paper. He was also being asked to admit that the Army did a better job than the Marines in developing a helmet for its troops—an absolute cardinal sin in the bureaucratic wars of the Pentagon.

Maybe, as Catto testified, he believed what he was told—that regular marines did not need a padded suspension system for their helmets. Ironically, Catto's boss, the Marine Corps commandant, and his aide had their padded helmets. Shortly before the hearing, a sales official from Oregon Aero and a Marine Corps brigadier general who was an aide to the commandant met at the Navy Annex in Washington DC. At the aide's request, the sales official gave him an Oregon Aero helmet-pad kit. According to the Oregon Aero official, the aide confided that even if the Marine Corps didn't field helmets with pads, the commandant and his aides have their pads and they will be safe. Also, the Oregon Aero official learned that the commandant of the Marine Corps wore an Oregon Aero pad kit in his helmet while visiting Afghanistan earlier in 2006.

"Could hardly see the General's face for all the tooth dust coming out of his painfully smiling mouth," Doc Bob recalled while listening to Catto's testimony. Doc Bob felt disgusted at what he was hearing. "As a helicopter pilot, he seemed to know little about what the ground troops (grunts) actually did in combat and why they would wear their helmets for hours or days on end," he felt. Doc Bob left the hearing room with the thought that Catto had embarrassed the Marine Corps "rather mightily in the hearing."[3]

The hearing may have been embarrassing for the Marine Corps, but it was a big success for Doc Bob and the efforts of Operation Helmet. Operation Helmet had now reached a new level of recognition and credibility both with Congress and the public. It was no longer just a battle between Doc Bob and the military with Congress now joining him in the effort.

After the hearing Doc Bob, with Cher in tow, spent the next two days appearing on C-SPAN, *Good Morning America*, and a number of local shows. They also took the opportunity to visit Walter Reed Army Hospital and Bethesda Naval Hospital, along with the Fisher House Foundation (a public-private partnership that provides lodging for the families of hospitalized military personnel; the houses are located at major military hospitals). Just Doc Bob and Cher. Nobody else. No media publicity. They just walked into the hospitals unannounced and visited the soldiers. According to Doc Bob, when they arrived at Bethesda and visited a marine with a brain injury and a leg gone, his father jumped up, ran right past Cher, and picked Doc Bob up, hugged him, and said, "My son is alive because of Operation Helmet." The father explained that his son had been injured when he was blown out of the turret of a Humvee and landed on his head. If he had worn the old PASGT helmet, he would have been dead, but he was alive. The grateful father's story was enough to bring tears to their eyes.[4]

Five days after the hearing, with the testimony still resonating in his mind, Representative Weldon signed a letter intended for Under Secretary of Defense for Acquisition, Technology and Logistics Kenneth Krieg, with a straightforward request: that new blunt-impact tests be conducted on padded suspensions systems. As an introduction to his request, Weldon referred to the complaints of marines and Operation Helmet's work in providing them with Oregon Aero helmet pads: "Apparently we have thousands of warfighters, primarily Marines, who believe that they are not being provided the necessary combat helmet protection they need from non-ballistic threats." Weldon wrote, "As a result, these men and women are depending on donated padded helmet suspension systems from a private charity, to give them the non-ballistic impact protection and comfort they desire. This is an unacceptable situation that has existed for some time, and needs immediate correction."[5]

By singling out the Marine Corps brass, Weldon let Krieg know that he was unhappy with the Corps's practice of not including a padded suspension system in its LWH, stating that it was "determined several years ago that an internal padded suspension system provides the best nonballistic blast and blunt-force impact protection." He continued by citing the Army, the Navy, the NFL, and the Department of Transportation as organizations that had already determined this. "Despite this," he said, "the Marine Corps insists that a sling suspension system best meets its needs."

Weldon then requested that Krieg "designate an independent test organization to immediately conduct all necessary ballistic and non-ballistic (to include blast and blunt-force impact) tests, using the same standardized testing criteria on the Light Weight Helmet used by the Marine Corps and the Army Advanced Combat Helmet. The Marine Corps helmet should be tested with the current sling suspension system and the Marine Corps and Army helmet should be tested using all four padded suspension systems used and qualified by the Army (Team Wendy, Skydex, Oregon Aero and Brock USA)." In addition Weldon asked that the results of the tests be "provided to the congressional defense committees and the Vice Chairman of the Joint Chiefs of Staff, with directions to the vcjs to immediately provide the best available combat helmet and suspension system, indicated by the test results, to our warfighters within 30 to 60 days."

CHAPTER 15

Small Victory, Large Defeat

While General Catto was defying Congress, claiming he would not authorize helmet pads for his regular marines, Lance Corporal Justin Meaders and other members of his unit doing "direct operations" were being hit by IEDs in Iraq. They were into their second deployment and were operating out of Al Asad Air Base in Al Anbar Province.

Two of Justin's friends were in an older, poorly armored Humvee that was hit by an IED. They were wearing the LWH with Oregon Aero pads, obtained from Doc Bob, and they survived without brain injuries. Their helmets stayed on, and they were able to get out of their vehicles and continue with the mission. Rockets and mortars came in very close, landing right on top of them, causing shock and disorientation, with potential brain injuries, but they survived pretty much unscathed.

Justin felt the effects of the explosions, especially during demolition operations, for he was as close as three feet from some of the explosions. He noticed a big difference in the blast effect he felt with a padded helmet versus a padless one. One of his friends, a gunner on top of a Humvee, was wearing a PASGT with a pad upgrade kit when he was blown off the Humvee by an IED. His helmet remained on his head, and he did not suffer a head injury.

Justin let his grandfather know that padded helmets worked amazingly well and offered significant protection and wearability. Doc Bob was determined to mainstream the helmet pads for the marines and was frustrated that he had to run a small nonprofit organization to get them what they needed. He could not fathom what was wrong with the Marine Corps equipment leadership.[1]

It is well known inside the Washington Beltway that the way to force an issue with the military is to embarrass it publicly. That was accomplished with the congressional hearing on padded helmets. It was embarrassing for General Catto to absurdly support the sling suspension system over a padded system despite solid evidence that the latter offered superior protection, something substantiated in his own Marine Corps's findings—test results that his subordinates either failed to provide him, or distorted, or simply ignored. The heavy demand by his own marines for the pads from the Cher-supported Operation Helmet was a continuing embarrassment over which he had little control.

The public embarrassment was a strong motivator that pushed a strongly resistant Marine Corps to act, even though only halfway at first. About two months after the hearing, the commandant of the Marine Corps issued a new policy that unit commanders were then authorized to use either the sling or the padded suspension system for the lwh. Then, finally, in October 2006 this order was followed with another that a padded suspension system "was now the only authorized suspension system to be used in the lwh."[2]

Along with this order, the Marine Corps purchased eighty-nine thousand sets of pad kits for immediate fielding for the lwh. Forced to see the light and accept the change, the helmet bureaucracy then admitted what the rest of the military already knew, that the padded suspension system offered more protection than the sling system.

However, in a thinly veiled shot at Oregon Aero and Operation Helmet, Marine Corps officials made it clear that the only helmet pad suspension system authorized for the lwh must be obtained from the supply system. Marines were not allowed to obtain pads from out-side sources. The primary pad kits in the Marine Corps's supply sys-tem at the time were from another pad manufacturer, Team Wendy, a lower cost but stiffer type of pad system. Doc Bob mused, "Typi-cal government move: mandate a good idea, i.e. pads, then require a lower quality pad system."[3]

At the time Doc Bob was not familiar with Team Wendy's helmet pads. Curious, he purchased a set of its pads directly from the com-pany. To find out their true value as wearable pads, he sent them to military veterans he knew who had been in combat and had worn the PASGT or the ACH. They reported back, likening the pads to bricks and suggesting they not be put into helmets. Not long after the supply-issued Team Wendy pads were fielded with the LWH in late 2003, Doc Bob started receiving emails from marines in combat, complaining that the pads caused much discomfort, including headaches, and distracted them from mission performance. There was a significant uptick of requests to Doc Bob for a change to Oregon Aero pad kits, mostly from marines, from approximately 8,700 requests in June 2006 to 35,000 requests in less than a year. According to Operation Helmet data, Marine Corps requests for different pads constituted 64 percent of all requests during the period, 92 percent of all requests during August and September 2006, alone.

When the Marine Corps issued an order requiring troops to exchange any pads obtained from outside the Marine Corps for the supply-issue pads, Doc Bob never knew of a single marine who followed the order. He asked some of the marines if they had to give up their pads. The response was usually that they didn't. In one case, according to Doc Bob, the commanding officer of a Marine unit told them, "Bullshit, keep what you have. I have them."[4]

Justin Meaders was in Iraq when the order came down that marines were to wear only pads issued from the supply chain. They were told they couldn't have pads obtained outside of what was issued. According to Justin, "We were told if we were ever killed and were not wearing the proper protective equipment we were supposed to, we would forfeit death benefits for their families." However, "we continued to wear the Oregon Aero pads and they never caught on." He never heard any officers say they had to change their pads.

Since he was receiving numerous complaints about the supply-issued pads, Doc Bob began exchanging emails with Team Wendy about its pads, sending the company an example of some of the soldier comfort complaints and asking if Team Wendy could fix the problem. But comfort was not its priority as former Team Wendy CEO John Sweeny said in an email to Doc Bob in July 2006: "We are working with Systems Command on an evaluation of our pads. I have heard some rumblings on the stiffness of our pads. They don't feel as soft as others. We optimize them for energy absorption with comfort as a secondary consideration."[5]

Team Wendy pads were not the only pads marines and soldiers were complaining about. The helmet manufacturer MSA Gallet also started producing its own pads after discontinuing its contract with Oregon Aero for the ACH and the LWH. The pads were also being purchased by the military and installed in the ACH and the LWH. Like Team Wendy MSA also sent Doc Bob a box of its pads. He, in turn, sent them to the same veterans who had tried out the Team Wendy pads. He asked them to put the pads in their helmets, sit around the house all day, and then tell him what they thought of them. According to Doc Bob, the gist of the return responses was, "Don't send those over; people won't wear them. You take off the helmet; you don't have a helmet on your head." Based on the fact that these troops knew which pads worked and which didn't work in combat, Doc Bob believed that they were probably right.

An Army soldier emailed Doc Bob and reported that the only way he could make his MSA pads comfortable was to hit them with a hammer until they turned into dust. Since he had MSA pads, Doc Bob also hit them with a hammer, and sure enough, they turned into granular dust inside. "It was like a bean bag," he said.

It wasn't long before the Army chief of staff contacted Doc Bob, proudly advising him that since the Army was fielding the ACH with pads, he did not have to send pads to soldiers anymore. This was good news for Doc Bob. Maybe he could now just go out and play golf. He had learned that the Army had five hundred thousand Oregon Aero pads just sitting in its supply system unused. Instead, it was buying new Team Wendy and MSA pads, at added expense, and sending those to the troops. This mystified Doc Bob. Why would the Army not use up what it already had in inventory? He would soon learn that Oregon Aero pads were no longer authorized. That seemed incredible to him. Thus requests to Operation Helmet for different pads increased. Doc Bob's golf bag was back in storage.

Doc Bob was distressed that there seemed to be a complete disconnect between the Army and Marine Corps's helmet bureaucracy decisions versus the reality of their troop's operational needs in combat with the LWH and ACH using supply-issued pads. He was aware of a conflict between the Army and Marine Corps and Oregon Aero, but he didn't know the extent of it. He made it his mission to find out what was going on.

Between December 2008 and July 2009, the Government Accountability Office (GAO) conducted a review of the Army and Marine Corps ground-combat helmet pads. Apparently accepting the word of Marine Corps and Army helmet officials, the GAO stated its awareness of "the use of unapproved pads by soldiers and marines," making an indirect reference to Operation Helmet as the culprit through which troops obtained the pads. However, "both the Army and the Marine Corps are aware of this problem and have issued directives specifically precluding the use of unapproved pads or other personal protective equipment."[6]

CHAPTER 16

Army Resistance

For Assistant U.S. Attorney David Peterson, there were two important factors in the Kevlar case that he wanted to demonstrate: did the amount of Kevlar woven into pattern sets for the PASGT helmets comply with what the contract called for and did the helmets pass ballistic testing? Technically, to prove that the contractor had made a false claim to the government, Peterson did not need to show that the helmets made with the Kevlar failed to pass ballistic tests. He only needed to show that the contractor had not complied with the specifications. However, for jury appeal ballistic test failures would be important.

It was also important to show in court that the Army had a legitimate reason for setting the weave density at thirty-five by thirty-five, that it was the minimum requirement to protect the soldiers. Peterson wanted to obtain records from the Army that documented the justification. Finding this documentation could be a problem, but he believed that Army investigators assigned to the case could track that down.

In the end Peterson felt that the most important criterion was the ballistic tests conducted on the helmets. According to the legal complaint, the helmets were supposed to pass certain ballistic tests. However, he needed more information about that claim and whether the helmets actually passed those tests. Sioux Manufacturing would later, after the case was unsealed, state in its answer to the complaint that the pattern sets and helmets had "never failed any of the ballistic test that [were] required as part of the contracts or military specifications applicable to these products."[1]

Peterson learned during the investigation that Sioux Manufacturing had a ballistics laboratory where it was supposed to conduct tests on the Kevlar pattern sets. In addition UNICOR sent sample helmets to an independent laboratory to conduct ballistic tests. The question still remained about the level of protection degradation caused by shorting the weave. Would the Army recall the helmets? Would it conduct its own ballistic tests? These were questions he needed to have answered, and he needed the Army's cooperation.

Notification of a potential safety issue involving the PASGT helmet should have been made to the Department of Defense after Campanelli submitted his complaint and disclosure statement to the DOJ on May 12. Once the notification was made, the DOD, especially the Army, should have immediately taken steps to determine the impact that the shorted Kevlar weave had on the helmets. However, there was no evidence that either a notification or an impact determination occurred. It was not until June 1 that an Army criminal investigator notified the Army at Natick and advised it of the shorted weave for the PASGT.

Even for the Army investigator, obtaining documents from an Army bureaucracy in support of a criminal investigation would be a difficult if not an impossible task. To the Army equipment bureaucracy, an investigative request for documents would be considered an intrusion in its day-to-day routine—one to avoid or put off. A notification that there may be a significant problem with a helmet presents a threat to the program, the funding, and maybe careers. Thus these bureaucracies vigorously seek to maintain the safe status quo.

Janet Ward, a materials specialist at Natick, brushed off the shorted weave by telling the Army investigator that it wasn't enough to degrade the ballistics integrity of the helmets. After all, she reasoned, until 1982 the standard was thirty-four by thirty-three. The bureaucracy at Natick and the developers of the PASGT helmet seemed to have a "so what" attitude concerning their requirements setting the minimum weave at thirty-five by thirty-five. Also examining PASGT helmets apparently was not a high priority. Ward told the investigator how difficult it would be to deconstruct a helmet to get to the individual Kevlar panels. Her lab and the lab at Aberdeen Proving Grounds could potentially deconstruct the helmet shell to get to the individual panels, but a fair amount of Kevlar thread would be destroyed in the process. They would have to go through several helmets before refining the deconstruction, possibly all the back stock of the lots in question, she explained to the investigator. Clearly it was a process the bureaucracy didn't want to go through.

Not surprisingly, Peterson was dissatisfied with Natick's response. "I didn't know what Natick's statement that the Kevlar was not degraded meant," he said. On August 2 Peterson and his investigative team teleconferenced with Ward to seek more information on the impact of a thirty-four-by-thirty-four weave on the PASGT helmet. Again, she downplayed the impact, rationalizing that, until 1982, the weave density standard had been much lower. It was becoming frustrating for Peterson. The Army, as part of the government, was supposed to be on the same side, but he was not getting anywhere with it. According to Peterson the Army didn't seem to be interested in responding to the problem.

It was not a matter of their being on the same side. Natick's initial response to Peterson, that nothing was wrong with its PASGT helmet, was not surprising. The military bureaucracy is largely made up of expensively funded programs that some bureaucrats will protect and hide mistakes at all costs. The Army helmet bureaucracy, like the Marine Corps helmet bureaucracy, is such a program; neither was willing to acknowledge a problem with its helmets. It was the same type of problem that Doc Bob was facing with the Marine Corps and that Oregon Aero encountered with the Army.

The Army designed and developed the PASGT helmet, a process that took seven years from design to production and involved great expense. Program managers had their government careers invested in the program with large buys of this helmet. It was still being produced despite the development of the new ACH. The ACH also used the Kevlar weave from the same contractors, had just gone into production, and there was a lot on the line for the people in the bureaucracy who had developed it. The Army was convinced that its new helmets were sound and free of defects. To admit there was a serious problem with a shorting of the Kevlar weave and possible protection degradation would open a real can of worms and present a serious threat to the program's funding. The Army did not want to explain the expense and hassle of a possible recall. Also careers would be at stake, promotions put on hold, congressional inquiries held, and more, which could upset the funding flow to the program the bureaucrats depended on for their livelihood. Beyond that the Army helmet bureaucracy at that time was trying to fend off the assault from Doc Bob and the issue of the padded suspension system.

A crucial question for the investigation was why the weave density was upgraded to thirty-five by thirty-five in 1982. Had testing determined that the earlier thirty-four-by-thirty-three standard was insufficient to protect the soldier? Ward's rationalization that the weave was lower until 1982 did not answer the question of why it was thirty-five by thirty-five in 2006. There had to have been testing that called for upgrading the minimum weave standard to thirty-five by thirty-five. Peterson needed more information on the history of the weave-density requirement. He had to get answers to his questions in order to obtain the necessary documentation, which called for a face-to-face meeting with Ward at Natick.

Peterson had to observe protocol to properly request an in-person interview at Natick. On August 23 he sent a letter to the commanding general of the Army Material Command, the authority over Natick, seeking the Army's assistance in the investigation and requesting that a meeting with Natick personnel be arranged. The Army responded by arranging for interviews at Natick.

On October 26 Peterson, his colleague Shon Hastings, and Ken Dieffenbach of the DOJ-OIG waded into the belly of the beast at the Natick Soldier Center to meet with Janet Ward. Ward was very knowledgeable about how Kevlar worked with the helmets and how Natick functioned as a research and development agency. However, Peterson did not get completely satisfactory answers to the question of why the weave density requirement was initially set lower than thirty-five by thirty-five and then later upgraded. He asked Ward if he could obtain the supporting documentation that had prompted the Army to set the weave density requirement at thirty-five by thirty-five. He was dumbfounded when she responded that the documentation could not be found.

Peterson, Hastings, and Dieffenbach left the meeting with very little accomplished, especially in getting the answers or documents Peterson wanted. All the signs indicated that the Army would resist cooperation with the investigation.

Andrew Campanelli could not believe what he was hearing when Hastings gave him the news that all of Natick's records regarding the initial testing of Kevlar and the manner in which they had set the minimum weave density had gone missing. The Army claimed that it didn't know how it had determined thirty-five by thirty-five to be the minimum, or know what ballistic testing was done to make that determination.[2]

It would be very damaging to the case if the DOJ couldn't obtain those records from the Army, Campanelli thought in frustration. If the DOJ couldn't do it, he asked himself, how was he going to do it if he had to take the case to trial himself? He recalled making a call to the Piget, Reed, and Johnson law firm in Mississippi when he was later in Fargo. He knew that Brad Piget had been the U.S. attorney in Mississippi under Clinton. He got one of the partners, Cliff Johnson, on the line. After explaining that the DOD was telling the DOJ that important documents had disappeared, he asked Johnson how to get those documents if Campanelli was forced to go up against the government. "I know there are certain regulations for getting documents from DOD. How does it work?" Campanelli asked. "Please educate me right now, right here, as I stand on the street." After thirty-five minutes of discussion about the problem, Campanelli realized the bottom line: if somebody high enough in the DOD wanted those documents to disappear, they were gone. That was the reality he had to deal with in the Pentagon bureaucracy.[3]

Then on December 22, 2006, an investigative team agent contacted an official (name not revealed) from the Army's PEO Soldier in the Washington DC area. It was a fairly new agency at the time, created as an adjunct to Natick to develop soldier equipment. In a moment of unusual candidness—for a government bureaucrat—the official admitted that the thirty-four-by-thirty-four weave resulted in a 3–4 percent reduction in ballistic protection. This admission contradicted Janet Ward's assurances that the armor would not be degraded. However, in February 2007, when this official was reinterviewed by the same agent, he backtracked, attempting to mitigate his previous revelation by saying that the weave shortage actually equated to a 2.7 percent loss of ballistic protection and stating that he did not think a recall would be warranted based on the shortage of one yarn.

At that point in the investigation, with the "disappearance" of the Army's important documentation supporting the decision to set the weave density at thirty-five by thirty-five, the Army was not going to recall the helmets—no matter that one Army official had admitted the lower density offered less ballistic protection—there had been no effort to perform additional ballistic tests of PASGT helmets. To Peterson and Hastings, this resistance by the Army presented a problem for their case to show fraud instead of just breach of contract.

On April 12, 2007, a meeting was convened in Washington DC between Peterson and DOD representatives to discuss the investigative findings. Peterson hoped to address important questions that the Army still had not answered after almost a year since it was first notified of the problem. Along with Peterson and the investigative team, DOJ attorneys, representatives from the DOD, officials from the U.S. Army's PEO Soldier, and UNICOR personnel attended the meeting.

During the meeting Peterson provided the DOD officials with a "to whom it may concern" memorandum containing a summary of the investigation and asking them to be prepared to answer a series of critical questions, with a response no later than May 12. It was a last-gasp effort to get information he had been trying to obtain from a reticent Army bureaucracy. At that time Peterson still did not have what he considered satisfactory explanations for why the Army had specified the thirty-five-by-thirty-five weave density, and the underlying documents had not been located. Based on the investigative findings, "it appears that until April 2006, SMC may not have ever complied with the 35 x 35 standard weave density in its construction of complete PASGT helmets or its manufacture of Kevlar helmet cloth for the pattern sets provided to UNICOR," Peterson wrote in his memo. "This practice potentially impacts an estimated 2,000,000 PASGT helmets."

The questions that Peterson believed were important to the resolution of the investigation included the impact of the less-dense Kevlar on the safety of PASGT helmet users, what actions the DOD planned to take, such as recalling the helmets, and any evidence of personnel having been injured or killed because the helmet did not provide the protection it was designed to provide. He also wanted to know why the Kevlar density specifications were set at thirty-five by thirty-five yarns per inch and whether DOD debarment of Sioux Manufacturing was a possibility.

Even though DOD officials had expressed concerns about the investigative findings, they revealed that PASGT helmets passed ballistic tests before they were fielded. Peterson felt that revelation seemed to give comfort to everyone at the meeting. Normally, the Kevlar panel sets went through ballistic tests at Sioux Manufacturing, and a private laboratory contracted by UNICOR tested the completed helmets. Army officials gave assurances that the Kevlar and the helmets passed those tests. However, DOD officials suggested that they were going to consider the issue more carefully by conducting additional ballistic testing and get back to Peterson.

By July 2007 Peterson had not received a response from the Army regarding the questions he had asked to have answered by May 12 or information about the additional ballistic testing the Army had promised to perform. He was determined to obtain the information he wanted, so he wrote another letter to an even higher authority, Secretary of the Army Pete Geren. He requested Geren's assistance in getting a response to his written questions of April 9. By raising the issue to the secretary level, Peterson hoped to finally get a response from the Army. This tactic seemed to work, but another month passed before he finally received a response, on August 7, from the DOD acting principal deputy general counsel, E. Scott Castle, with an enclosed response to Peterson's questions by PEO Soldier. It took four months with considerable effort on Peterson's part to access the information. Remarkably, PEO Soldier revealed that Natick had suddenly "identified" the "lost" historical files regarding the development of the weave-density specifications and said the documents would be sent to Peterson.

The Army was no longer buying PASGT helmets, but in 2006–7 thousands of those helmets were still being used by soldiers and marines in Iraq and Afghanistan. Castle revealed in his cover letter that the ballistic tests conducted by the director of Operational Test and Evaluation (DOT&E) in July 2007 "verified that the PASGT helmets met U.S. Army performance requirements." However, there was no additional information about what test methods were used, the origin of the helmets used for the tests, how old the helmets were, or identification of the helmet manufacturer and the contractor that had woven the Kevlar cloth.

It was also revealed that no additional actions, including a recall, were anticipated. PEO Soldier claimed that the Army was not aware of any "instance of injury or death caused by a failure of a PASGT helmet to perform as designed." The only action the Army decided to take was to issue a directive on July 27, 2007, mandating the development of a uniform standard of ballistic protection among all service components using the ACH and the Marine Corps's LWH.

Additionally, the Army informed Peterson in its response that the Defense Logistics Agency makes an initial decision to recommend debarment of a contractor, but the final decision resides with the Army's program contracting officer. As far as his case was concerned, Peterson felt the most significant issue was that the helmets apparently passed ballistic tests.

Despite evidence of underweaving based on the results of all the records seized and the interviews, Peterson decided to decline the criminal part of the case because he felt he could not prove criminal intent. He was looking at it from the perspective of someone who had tried some two hundred jury cases during his career. Because of the Army's actions described in Castle's letter, there was no sure way to prove criminal intent. Based on poor information the Army had provided Peterson, the alleged positive PASGT helmet test results, and the contention of Army helmet officials who would testify that no harm resulted from shorting the Kevlar weave, the case boiled down to a breach of contract issue. Peterson made this decision despite the fact that the government had not received the amount of Kevlar it had contracted and paid for.

During early August 2007, Peterson received a call from Army officials informing him that some of the records he had requested regarding the weave density were being shipped to his office. More than a year after first requesting the documents and having eventually appealed to the top of the Army food chain, Peterson and his colleagues finally received a box containing a lot of duplicate material, a lot of junk, but also some information they were looking for. The records revealed that until 1982 the weave density requirement for Type II Kevlar cloth was thirty-four by thirty-three yarns per inch. However, in July 1982, Kevlar manufacturers reported to the Army that Type II cloth, woven under the specifications at that time, were "barely meeting the minimum ballistics requirement."

This information, although not entirely clear, concerned the Army helmet bureaucracy because it indicated there might have been a trend toward development of lower ballistics protection for helmets at the weight and density set for the cloth. However, it was clear to the Army that fabric suppliers would "design toward or at the minimum weight required by subject specification. This could result in large procurements programmed for the next five years with marginal ballistic properties." The Army felt it was a "safe assumption" that due to the high cost of Kevlar, suppliers would "continue to produce fabric at the minimum weight and gamble with the ballistics."

As a result Army officials decided to raise the minimum weight and density requirement to thirty-five by thirty-four yarns per inch to better achieve acceptable ballistic performance. On September 24, 1982, the cloth specification was officially implemented, raising the weave density and the weight requirement.

In June 1986 the weave density requirement of the Kevlar cloth was again raised, this time to thirty-five by thirty-five yarns, or picks, per inch. The Army intended the change to "ensure acceptable ballistic protection using commercially available Aramid fiber which, by DuPont's admission, ha[d] a plus or minus 90 denier variation." Additionally, Army officials pointed out that casualty reduction studies had not been conducted on Type II Kevlar fabric having less than the minimum number of yarns per inch.

Soon after the new density requirement was set, an unknown Kevlar cloth supplier requested a waiver of the thirty-five-by-thirty-five requirement. Army officials denied the request, noting that due to the "life and limb nature of this item, no material [could] be accepted that ha[d] not passed a casualty reduction study." No wonder the Army was reluctant to find these documents because it showed that what Sioux Manufacturing was doing was threatening the "life and limb" of the troops. This documentation confirmed that Tammy's and Jeff's instincts were right.

CHAPTER 17

Cut and Paste

D espite a congressional mandate in June 2006 requiring the DOD to conduct all-new helmet pad testing, the result was a deception. On February 22, 2007, DOD Undersecretary Kenneth Krieg responded to chairman Kurt Weldon of the House Subcommittee on Tactical Air and Land Forces with an executive summary accompanied by a detailed technical memorandum announcing that "all necessary non-ballistic, blunt impact testing [was] complete."[1] However, the testing reportedly conducted either did not occur or appeared to be rigged.

In the world of journalism, one of the highest levels of the field is the investigative journalist, who has to ignore the day-to-day noise of the daily news and do the deep dive into complex problems that government and corporations do not want exposed to the public. Many of these reporters still work as freelancers so that they can dig deeper into topics that are important but not trendy, and they are often the only ones who can bring attention to corporate and government wrongdoing that otherwise gets overlooked.

Just north of downtown Washington DC, there sits a large complex of apartments near the Metro's Red Line, which shuttles thousands of its resident commuters into downtown and back on a daily basis. One of those residents is the prize-winning, longtime freelance investigative journalist, feature writer, and author Art Levine, who has written a wide range of investigative pieces and features for the *Washington Monthly*, *Salon*, the *American Prospect*, *Mother Jones*, the *Daily Beast*, *Slate*, *AlterNet*, the *New Republic*, the *Atlantic*, and other leading news outlets. He was honored in 2001 as Journalist of the Year by the Florida chapter of the National Alliance for the

Mentally Ill for his articles examining the criminalization of the mentally ill in South Florida. His book, *Mental Health Inc.: How Corruption, Lax Oversight and Failed Reforms Endanger Our Most Vulnerable Citizens*, exposes America's broken mental health system. He also worked with producers for Dan Rather for a piece on the helmet-pad issue.[2]

In 2007 Levine, known for his hard charging and obsessiveness in his journalistic endeavors, became interested in the helmet-pad controversy through Doc Bob and Operation Helmet. As a result he began a long odyssey researching how the DOD rigged equipment studies in a way that raised the risk of brain injuries to the troops in order to force out a vendor, Oregon Aero, which had been providing safe helmet-padding equipment for years, and bring in a less expensive manufacturer. Levine also found the controversy surrounding Natick and Oregon Aero intriguing and instructive on how bureaucracies can fail the people they are supposed to be helping.

A major focus for Levine was the 2006 congressional mandate for new blunt-impact tests and what the DOD claimed to be the mandated-test results. His findings seemed to indicate falsified studies using dubious data and an apparent bureaucratic scheme to oust Oregon Aero, at a time when thousands of soldiers and marines were experiencing TBI and needed both head protection and comfort. Levine was doggedly determined to know what exactly took place and why.

The executive summary prepared by the Office of the Secretary of Defense and sent to Congress in February 2007 seemed at first glance to indicate that the results were derived from all-new tests, as mandated, conducted in 2006 by the USAARL: "The tests were performed by the U.S. Army Aeromedical Research Laboratory, which had conducted similar tests on the ACH in the past," the report claimed. However, Levine's careful reading of the accompanying technical memorandum, prepared by USAARL, indicated that most results were not based on new contemporaneous tests but cut and pasted from an earlier USAARL test report, a suspicion that was later confirmed by current and former DOD test officials. Levine noted that the earlier dubious report was officially known as USAARL Report 2005-12, issued in August 2005. That careless test report did not identify the manufacturer, including Oregon Aero, of the pads that were used in those tests. Some additional tests were apparently conducted in 2006 that included Team Wendy pads and a different version of Oregon Aero pads not tested for the 2005-12 report.

The 2005-12 test report was the one previously noted in chapter 9, which Natick's George Schultheiss had revealed to Oregon Aero in February 2006, claiming the company had failed performance tests by scoring 278 g's in the hot-temperature tests. Also Mike Dennis of Oregon Aero said, as stated in chapter 9, that his pads were not used in the testing based on how his pads are submitted for testing and the use of certain identification markings.

Levine alerted a congressional aide from the House subcommittee about the falsity in the executive report and the technical memorandum Congress had received from the DOD. At an isolated corner table in a congressional lunchroom, Levine met with the aide and walked him through the executive report, the technical memorandum, and USAARL Report 2005-12 while leaning over the aide's shoulder and using his finger to point out how the information from the USAARL report was cut and pasted into the new report to Congress. The aide was very surprised because he had been completely unaware of the fakery. He had assumed that the reported test results were from newly conducted blunt-impact tests conducted contemporaneously. He was outraged.

The congressional aide told Levine that the whole purpose of mandating new tests was to establish a new baseline by using all qualified systems for all helmets. It would have the result of wiping the slate clean to address any problems vendors had previously with the Army. The aide was adamant that the letter to Undersecretary Krieg was straightforward and clear. It required the DOD to conduct a new round of blunt-impact tests to qualify the helmet pad systems because Oregon Aero's pads, with a history of being used by the Army, were being called into question. It was important for the committee to determine the validity and fairness of the previous tests.

As a result of the false information contained in the report submitted to Congress, the technical memorandum had concluded that the Oregon Aero pad system provided for the Marine Corps's LWH did "not meet the Army ACH performance requirements for blunt impact protection when tested at the 10 fps impact velocity." These "tests" would later be used to disqualify Oregon Aero based on the purported 278 g score for the "hot" test, which did not reflect actual operating conditions in the field, and allegedly to justify purchasing other, less expensive pads.

If the military's far-fetched claims are to be believed, Oregon Aero suddenly started manufacturing shoddy pads for the ACH and the MICH in 2003 despite standards they had easily met or exceeded since 1999. If that were true, Levine wondered, why wasn't there even one email, letter, announcement, phone call from anyone in the military or the helmet manufacturer in 2003, or even 2004, pointing out the company had failed 2003 blunt-impact tests? Why was there no correspondence between the Army and Oregon Aero in 2003 claiming that the company no longer met its original helmet standard, which it had achieved in 1999? Why did it take until 2006, three years later, to reveal the issue? If Oregon Aero had actually failed tests in 2003, it would have been the duty of the Army to address the problem by immediately notifying Oregon Aero. Actually, no results ever showed that Oregon Aero pads had flunked performance tests in 2003. Instead, there was just one test conducted in 2003 by Natick Soldier Systems labs showing Oregon Aero pads fared very well, exceeding the 150 g standards—even at a more demanding rate of impact.

Even more baffling and highly unlikely was the statement in the technical memorandum that most of the tests in Report 2005-12 report were conducted in 2003, at a time when Oregon Aero was the sole or primary provider of pads to the Army and Team Wendy had not begun producing helmet pads for the ACH. But the 2005-12 report wasn't published until August 2005, two years after the purported tests were conducted. Therefore, it was unclear where the 2003 test date originated. Levine felt that was an especially critical point because Oregon Aero wrote and later met with USAARL staff in early 2006 to say that the results in the 2005-12 report were not from a test of its product—long before the congressionally required USAARL report was completed in December 2006. In addition Levine noted that non-ballistic impact tests, conducted by both independent and military organizations between 1999 and 2004 for the Special Forces MICH and the new ACH (using Oregon Aero pads at the time), showed that their pad systems easily passed the military's 150 g specification with averages less than 100 g's

Besides the controversial hot-temperature tests and the "no-name" test report, Levine noted another anomaly. Head-to-head comparisons of the purported additional tests were conducted in 2006 between the one-half-inch-thick Oregon Aero pad kit and three-fourths-inch-thick pad kits from other manufacturers, including Team Wendy, that obviously resulted in a poorer performance by the Oregon Aero pads. It was, to Levine, apples-versus-oranges testing.

The director of Operational Test and Evaluation (DOT&E) represents the top testing agency within the DOD. Established by Congress in 1983, the director is the principal staff assistant to the secretary of defense on operational test and evaluation in the DOD. This official is responsible for oversight of important weapon systems and equipment testing to ensure it will operate safely and effectively in combat. Before a weapon system goes into full production, the DOT&E has to certify to Congress and the secretary of defense that the system has proved to be operational, effective, and operationally suitable, and the testing conducted was adequate for those conclusions to be drawn.

DOT&E had no control or authority over the mandated testing of the helmet pad or the 2006 report to Congress. However, the director's office did review the report and made some comments before it was released to Congress. Through a network of contacts and hard work, Levine developed a number of sources, two of which were former officials with DOT&E. One source, after reviewing the report, felt that the tests did not meet his standards. The source believed that the testing lacked critical information, including dates, location, times, and types of tests. He pointed out that test reports normally include a scope paragraph that states those elements.

One former senior DOD testing official attributed the lack of technical competence to the military equipment agencies, including the laboratories and the development agencies whose budgets had been gutted for years, the result of decisions that they didn't need such extensive capability. Congress had been pushing to cut back on their staff and the people overseeing them since the 1980s; the push accelerated in the 1990s. The source said that the budget cuts ultimately reduced the technical competence in the military departments, which still had a lot of deadwood after many of the sharp people had left.

Another source, a former DOD official familiar with the execution of the 2006 congressional mandate, told Levine that in order to implement the mandate, the DOD first pulled together a team to get the tests executed. The team designed the tests by reproducing prior testing reported in the now-controversial Report 2005-12. The team members agreed that the data was sufficient to include as part of the 2006 effort. As a result the team was allowed to reduce the scope of the 2006 test because the 2005 data could be used and the new report could be completed less expensively and more quickly.

Thus in the opinion of this source, it was clear that the team's approach taken in the 2006 effort would have used previous ballistic and nonballistic data because the team felt that this data satisfied some of the requirements of the 2006 effort. Based on that, the source believed it would have saved taxpayer dollars. However, it was not apparent to a reader of the report that old test data was being used.

A former DOD procurement official with knowledge of PEO Soldier revealed to Levine that the agency edited USAARL documents and pressured its engineers to come up with predetermined conclusions and omissions; this source told Levine that "the fix was in" and that "tests were rigged." The source believed that the DOD/USAARL had deceived the House Armed Services Committee. Also the source claimed that the technical report was edited with "malice and aforethought" by omitting all references to the MICH and the earlier 150 g standards met by Oregon Aero.

Philip Coyle is a retired director of DOT&E. Prior to serving as director, Coyle served in the Carter administration as principal deputy assistant secretary for defense programs in the Department of Energy, was deputy associate director of the Laser Program at Lawrence Livermore Laboratory and was nominated by President Obama for the post of associate director for national security and international affairs, office of Science and Technology Policy. Coyle expressed to Levine his concern about the tests reported in the 2005-12 report and the additional tests conducted for the congressional report. What Coyle read in the reports looked like "they ha[d] been summarized. That's where mischief comes in sometimes, is when people summarize raw data." He believed that the DOD's reporting to Congress that all new tests were conducted as mandated in 2006 was a deception.

"The test, to say the least, was disingenuous," former high-level DOD engineer and testing expert Pierre Sprey told Levine. "The only new thing they did for the Congress, it was very hard to dig out of the test report, they obviously buried it, they didn't want the Congress to know they hadn't done very much. . . . It wasn't what the Congress wanted to see."

"They tested the various kits that were available for improving the Marine helmet. For the Army helmet, the only new thing they did was to test the helmet at 14 ft./sec., at the higher level they had already concluded that none of the pads could meet. Everything else was ancient data. In particular, they are still leaning on this old data from 2003 that was done before testing of competitors in 2005 and showed enormously high levels for Oregon Aero," Sprey added.

"Now, they do this set of tests, unpublished in any separate report, later given to Congress, where they test the competitors, which were Team Wendy, Skydex, MSA, and Gentek," Sprey continued. "In those tests, they played out, the only case they could make against Oregon Aero is at hot temperatures. They are testing at cold temperature, somewhere below freezing, at ambient temperature, which is room temperature, and they are testing at 130 degrees Fahrenheit. This, in itself, is an astonishing piece of incompetence and germane on how they pushed Oregon Aero out of business. This is crucial, the hot tests, the insistence of doing tests at 130 degrees which is ludicrous from the point of view of combat. That's the thing they used to push Oregon Aero out."

Levine decided to go directly to the source of the technical memorandum, B. Joseph McEntire, an engineer with USAARL and author of the 2005-12 report. In a written response to Levine's submitted questions, McEntire confirmed that the tests reported in the 2005-12 report were conducted in 2003. His explanation for the no-named pads used in the tests was that pad vendor names were intentionally omitted from the report to "increase the report's distribution." His reasoning, difficult as it was to understand, was that "inclusion of vendor names could have caused the report to be considered a Test and Evaluation (T&E) report that would have limited its distribution to US Government and its contractors only. Thus, the pads were indeed Oregon Aero pads and at the time of the 2006 helmet and pad testing, it was not deemed prudent to expend Government resources retesting the Oregon Aero pads at the 10 fps impact velocity." He did not mention if he believed that the pads tested in 2003 and reported in the 2005 report were Oregon Aero.

Levine didn't know whether McEntire was a pawn of Natick or was just sloppy. USAARL even sent Oregon Aero photos of the helmet with the pads that were tested. Based on what Mike Dennis saw in those photos, he was certain that the pads were not Oregon Aero pads. He noted that the pads Natick had sent USAARL were unmarked, and Oregon Aero never produced or sold unmarked pads.

A high-ranking DOD testing official told Levine that an internal DOD inquiry did not turn up any documentary evidence that the Oregon Aero pads were tested for the 2005-12 report. Also, after comparing the 2005-12 report with another, similar USAARL report (2005-05) that was produced the same year, Levine was confident that the 2005-12 report did not follow standard reporting procedures. Unlike the 2005-12 "no name" report, the other report clearly identified the articles tested by product name and manufacturer with photographs of the test articles clearly showing the product label. Levine also noted that the Army continued to use Oregon Aero pads in 2003-04 for both the MICH and the ACH.

Although there was no documentary or physical proof that Oregon Aero pads were actually used for the tests reported in USAARL's 2005-12 report, circumstantial evidence strongly indicated that those pads were not used. Also, Levine's comprehensive investigation raised doubts that the tests reported in 2005-12 alleging that Oregon Aero pads failed the hot-temperature tests were not actually conducted in 2003. There were two important reasons for this suspicion: another lab test by Natick conducted in 2003 showed that Oregon Aero easily passed the 150 g requirement; and no military agency ever reported to Oregon Aero that it had flunked a test in 2003 or that its performance standards for the MICH had changed for the worse since 1999.

From Levine's investigative findings, he concluded, "The U.S. Army Aeromedical Research Lab (USAARL) jiggered the results of a 2005 study on an unnamed helmet pad system, 'finding' that those pads failed a recommended Army blunt trauma standard of 150 g that OA [Oregon Aero] had long met. Those failing test results were then falsely blamed on Oregon Aero in a 2006 USAARL report to Congress used to disqualify OA pads."

It may be easy for the Army to claim that the helmet-pad testing controversy was an anomaly, a one-time incident, but that is not accurate. Questions surrounding the Army's test procedures of important protective equipment to either favor or eliminate certain vendors were not limited to helmet-pad testing. Around the same time the helmet-pad testing controversy was drawing attention, Army testing and procurement officials were accused of rigging tests of body armor to favor a vendor. The controversy surrounding the body-armor testing actually started when Pinnacle Armor Inc., makers of a body armor called Dragon Skin, accused Army officials of unfairly favoring a rival body-armor supplier, Point Blank, which provided Interceptor Body Armor (IBA) to the military. Pinnacle claimed that the Army's tests of Dragon Skin conducted from August 2005 to June 2006 were altered to favor Point Blank. Those tests resulted in a number of complete penetrations of the Dragon Skin body armor, which Pinnacle officials felt were fraudulently arranged.

NBC News picked up the controversy in 2007 and arranged for independent ballistic testing comparing the two body-armor designs that were overseen by former DOT&E chief Philip Coyle. Later testifying before the House Armed Services Committee at a June 2007 hearing looking into the body-armor controversy, Coyle revealed that "the results from these limited tests favor[ed] Dragon Skin over the current military Interceptor Body Armor." Not surprisingly, Coyle concluded, "[The] tests contradict the information provided to this committee by military and Department of Defense."[3]

Based on his observations of the ballistic tests in Germany, Coyle believed that the Dragon Skin armor offered soldiers more advantages than the Army-favored Interceptor armor. "Dragon Skin performed perfectly," Coyle told the hearing committee members, "allowing no penetrations, and defeated six rounds of a particularly deadly ammunition threat which U.S. troops in Iraq and Afghanistan may face." In addition Coyle believed that Dragon Skin armor provided better protection against multiple shots and reduced blunt-force trauma—also important to the soldiers in combat.

Ignoring the evidence that Dragon Skin was the better armor, the Army equipment bureaucracy continued to prefer IBA. Despite the Army's insistence and manipulations to favor IBA, the online news outlet *Defense Review* reported in a 2008 article covering the body-armor issue that the vice chief of staff for the Army at the time, Lt. Gen. Peter Chiarelli, had "ordered eight sets of Dragon Skin for himself and his personal staff in May 2006 while serving in Baghdad.... Dragon Skin may not be 'good enough' for today's Soldiers, but it was fine for Lt. Gen. Chiarelli and his horse-holders."[4]

In October 2009 the Government Accountability Office issued a report also criticizing the Army for not following established testing protocols on a new design of body armor tested in November–December 2008. The GAO found that the Army again improperly scored test results indicating a complete penetration of a projectile as a partial penetration, which resulted in some armor passing when it had actually failed. The GAO stated, "It is questionable whether the Army met its First Article Testing objectives of ensuring that armor designs fully met Army's requirements before the armor is purchased and used in the field."[5]

Historically, there were several infamous examples of the Army equipment bureaucracy favoring vendors who produced combat equipment that was injurious, with sometimes fatal consequences, for soldiers in combat, over vendors who could produce a higher quality and safer product. An egregious example of Army bureaucratic intransigence occurred during the Vietnam War with the M-16 rifle. In the late 1950s, the Army designed a rifle that became the M-14, a rather heavy, unwieldy weapon that fired .30-caliber ball powder ammunition and was almost uncontrollable during automatic fire.

Not long afterward, during the early stages of the war in Vietnam, an outside designer named Gene Stoner developed a lighter automatic rifle called the AR-15 that used a smaller caliber ammunition and was easier to use in combat. Because of a complicated set of circumstances, several thousand AR-15s were deployed to Vietnam primarily for use by Special Forces operatives, and they performed brilliantly, vastly better than the Army's preferred M-14. However, the M-14 was the product of the Army rifle bureaucracy, and just like the helmet bureaucracy, it resisted the change to the AR-15 because it came from an outsider and presented a threat to the bureaucracy's M-14 program. Also the bureaucracy had a very cozy relationship with the vendor who manufactured the rifle and the vendor who produced the .30-caliber ammunition.

Due to enormous political pressure, the rifle bureaucracy relented and permitted the use of the AR-15 design, but decided to militarize it into what became the M-16. However, this militarization included several modifications that Stoner declared would make the M-16 unreliable. Also, the rifle bureaucracy conducted its own tests comparing the AR-15 to the M-14, which naturally resulted in the assessment that the M-14 was the better rifle and the recommendation to reject the AR-15. However, according to a report by James Fallows in the July 1981 issue of the *Atlantic*, an Army inspector general investigation "found that the tests had been blatantly rigged." Fallows recounted the full story of the AR-15 and M-16 rifles in great detail in his book *National Defense*.[6]

According to defense analyst Pierre Sprey, "What they did to combat the threat of the AR-15 was to deliberately doctor the ammunition of the M-16, for fielding to the Army, so the rifle would jam." Sprey explained,

It was so bad the manufacturer, Colt, producing the rifle, wrote the Army saying they had to stop production because they could not meet production tests for reliability because the ammunition the Army was sending to them was jamming. So they were going to stop production until the problem was solved. The Army rifle bureaucracy wrote back that it was ok they could use the old ammunition, but they were sending the new ammunition into combat. They were condemning thousands of Americans to death with jammed rifles in their hands in order to defend a bad rifle. Congressional pressure brought on by numerous letters any families of soldiers in Vietnam complaining about the jamming of the m-16, resulted in a House Armed Services Committee investigation that produced what was called the Ichord Report. The report concluded that the actions of the Army with respect to sabotaging the m-16 bordered on criminal malfeasance.[7]

In an article titled "Gun Trouble" that appeared in the January–February 2015 issue of the *Atlantic*, Robert Scales wrote of his and his soldiers' troubles with the m-16 during the Vietnam War, which resulted in the deaths of some of his troops during battles:

> With a few modifications, the weapon that killed my soldiers almost 50 years ago is killing our soldiers today in Afghanistan. . . . During my 35 years in the Army, it became clear to me that from Gettysburg to Hamburger Hill to the streets of Baghdad, the American penchant for arming troops with lousy rifles has been responsible for a staggering number of unnecessary deaths. Over the next few decades, the Department of Defense will spend more than $1 trillion on f-35 stealth fighter jets that after nearly 10 years of testing have yet to be deployed to a single combat zone. But bad rifles are in soldiers' hands in every combat zone.[8]

Thus military bureaucratic rejection of better-quality combat equipment designed by outsiders has been a pattern for generations. In order to protect its turf, the bureaucracy has been reluctant to accept change or innovation and to adapt to combat needs in war—even at the expense of the soldiers. This self-perpetuating power is rooted in a military bureaucratic culture that leaders such as secretaries of defense, Congress, and presidents have failed to alter. Most disheartening, because soldiers die in war, the military bureaucracy's malfeasance in these weapons failures is not as obvious as a car manufacturers' culpability for a bad design that results in driver fatalities. There are many ways to die on a battlefield, so it is hard to pinpoint or prove that a badly designed weapon caused a death. It is even harder for U.S. civilians to believe that the military would put careers and egos above soldiers' lives. In a similar fashion, the DOD was deceiving Congress about helmet-pad testing at the same time that thousands of troops in Iraq and Afghanistan were being exposed to blasts while wearing questionable helmets, which often led to TBI.

CHAPTER 18

Damage Control

If you must sin, sin against God, not against the bureaucracy. God may forgive
you, but the bureaucracy never will.

—Adm. Hyman Rickover, USN

To the Army's Aeromedical Research Laboratory, the USAARL, Doc Bob was the devil. He gave them fits with his public criticism of the testing agency. It was said that lab staff couldn't utter his name without cursing at the same time. This animus stemmed from Doc Bob's critical remarks, through his Operation Helmet website, about USAARL's tests, specifically the helmet blunt-impact studies with padded suspension systems. Also, to the chagrin of both USAARL and the military equipment bureaucracy, Doc Bob had gained a new level of respect from his appearance before the congressional committee. He was neither timid nor deterred in weighing in on the laboratory's technical memorandum submitted to Congress in 2006.[1]

In March 2007 Doc Bob countered the memorandum with a letter to Undersecretary Krieg and a copy to the House Armed Services Committee. While limiting his discussion to the LWH and PASGT helmet, he let it be known that no one at the USAARL ever contacted Operation Helmet to ask questions or confirm the facts that each helmet, specifically the LWH and the PASGT, required a different pad system.

"They [USAARL] used faulty reasoning," Doc Bob wrote, referring to the technical memorandum, "used bad end points for their data, and their data was inconsistent statistically, they couldn't defend how they came to their conclusions and they couldn't tell him why they chose those points. Why did they test at 135 degrees of the human head and a helmet on it? A human head is dead at 135 degrees. They tested the helmet and the head both at 135 degrees Fahrenheit. Why? They said that was what they were told to do."

Critical of pad placement on the test form, Doc Bob told Krieg that it did not "take into consideration the manufacturer's recommendations for dispersal of pads in the helmet to mitigate blast/ impact protection in the IED environment which is not mentioned in the report." Commenting further, he said, "This is perhaps the most important laboratory test, after blunt impact testing and was not included for reasons unknown to us. In the real world, pads are worn continuously in hot environmental atmospheres where sweat production is ongoing and pads are in danger of becoming waterlogged. An appropriate test should be designed in which the pads are immersed in water and subjected to continuous pressure equal to helmet weight on the wearer's head for a time equal to most combat patrols, then tested immediately."

Doc Bob's letter to Undersecretary Krieg may have caused concern within the House Armed Services Committee. Soon after the letter was received, House Armed Services Committee aides began raising questions with the Army about the validity of the tests reported to the committee. "There are concerns regarding the validity of congressionally mandated tests conducted last summer," the House Armed Services Committee stated in its fiscal year 2008 Defense Funding Bill submission. "The Committee feels this equipment is too valuable to our service members for it not to have a comprehensive test and evaluation including an operational evaluation of all systems by qualified military personnel under realistic environmental conditions." As a result "the Committee is directing DOD to retest all available helmet pad suspension systems."[2]

Now it was damage-control time for the Army and the Marine Corps. Army officials responded by requesting a meeting with committee aides in an attempt to "brief" them on the validity of the tests. One of the aides later told Art Levine that they felt the Army's request for the meeting seemed "fishy" but agreed to meet with the officials.

The meeting took place on June 19, 2007, between committee aides and officials from Natick, PEO Soldier, USAARL, and the Marine Corps. Levine learned, through a source familiar with the meeting, that instead of debating the merits of the tests as planned, the Army reframed the discussion to a forum on the merits of excluding Oregon Aero as a viable vendor.

The military officials claimed to the congressional aides that they had conducted operational testing of the ACH and the LWH but did not mention that some tests were not new and allegedly conducted in 2003. At the time of the meeting, the aides had no idea that older tests were cut and pasted into the report and passed off as new tests.

According to a briefing paper of the meeting, obtained by Levine, the results from those tests were used to make procurement and/or fielding decisions for the helmets pads. Using confusing military jargon and technical language, the military officials did not directly imply that the tests of Oregon Aero pads were conducted in 2003. Instead, they alluded to the use of previous tests, telling the aides that at the time of the operational assessment of the ACH, only Oregon Aero pads were fielded and therefore it was the only system tested.

Anticipating possible questions from committee aides regarding Doc Bob's critique of the tests, the military officials claimed that both services were satisfied with the performance of their helmet systems, and thus no further operational tests were planned. They did admit that in user surveys Oregon Aero pads were rated as more acceptable than Team Wendy pads regarding comfort in very warm weather, hot spots, pressure points, and skin irritation. However, in qualifying those results, they noted that the differences were not significant enough to justify further tests on fielded helmets since they had already been proved effective.

In order to convince the aides that all the tests were new, Army officials produced shipping receipt documents, dated during the summer of 2006, which allegedly verified they had received Oregon Aero kits for testing. They did not produce shipping receipts for Oregon Aero dated in 2003.

The Marine Corps official weighed in, telling the aides that the

Corps had not received any adverse comments regarding its helmet similar to what Doc Bob had reported. He justified the stiffer pads by claiming that the softer Oregon Aero pads, although more comfortable, could decrease impact protection, but he did not produce any data to prove his assertion. The slow roll on the congressional aides was complete.

CHAPTER 19

Negotiation

I n addition to the problems experienced by Assistant U.S. Attorney David Peterson in prying information out of the Army, the case against Sioux Manufacturing had become a political hot-button issue. The Spirit Lake Reservation was a distressed area of North Dakota, and the Native American people were having a tough time making a living. It was a sensitive social issue where the DOJ had to tread lightly. Unemployment rates on the reservation typically ran about 50–70 percent, and the North Dakota congressional delegation had tried very hard to get industry and development on the reservation and employment for its people.[1]

Despite the depressed state of most of the people on the reservation, Sioux Manufacturing received a record $30.2 million in revenue during the 2006 fiscal year and would post another record $32 million in revenue for the 2007 fiscal year. Peterson wasn't normally concerned about political issues. He cared only about the lawsuit— whether the allegations were true and provable. That was his job and focus. However, Sioux Manufacturing's retained counsel Sarah Vogel did have political connections and pushed enough political buttons that Peterson had to be concerned. He recalled that Vogel was well connected in the state as she had been a North Dakota state agriculture commissioner for two terms, and her father was a former state supreme court justice.

In August 2007, a little more than a year after the Sioux Manufacturing case was filed, Peterson moved to settle the case. The investigation had been lightning fast for a *qui tam* case. The DOJ takes three to five years, on average, to investigate any complicated DOD procurement case. Going into settlement negotiations, Peterson had wrestled with the amount of damages to the government. The issue was debated extensively within the U.S. attorney's office but it was difficult to come up with a value. What would be the value of the helmets? Is it the whole value or a partial value such as the value of the weave? Or is it the value of the entire contract or a portion of it? From a realistic standpoint, how much money would the Spirit Lake Tribe and Sioux Manufacturing have to pay on the financial settlement?

The evidence showed that the Kevlar cloth woven by Sioux Manufacturing did not meet contract specifications. The shortage ranged between 2.8 and 5 percent, which would result in damages to the government in the range of $1.4 million to $2.5 million based on total contract payments of approximately $52 million. However, the investigation could not estimate how often the Kevlar cloth produced under the contract fell short. Peterson knew that because of the circumstances, case law would not support using this method of calculating damages.

Also, by law the court was allowed to award penalties for each of the 207 claims in the range of $5,500 to $11,000 per claim, which in this instance would amount to a potential $1.3 million to $2.27 million. Again, the investigation was unable to prove that all 207 claims were false and thus was unable to come to agreement on the damages. Another factor that complicated a settlement amount was that the helmets had apparently passed ballistics tests.

Peterson also had to take into account that the only money the tribe had was appropriated to it by the government or through government contracts. That factor always came into play. The $150 million demanded in the complaint was not realistic under the circumstances. Despite whatever value Peterson placed on the case, the bottom line was that the DOJ was the one that made the ultimate decision regarding an acceptable settlement.

After considering all factors and with extensive debate with his colleagues, Peterson decided to start the settlement negotiations at $7.6 million with a bottom line of $5 million. He called Campanelli and discussed his settlement decision with him.

Initially, Campanelli thought that Sioux Manufacturing had been paid over $100 million on the helmet contracts, which formed the basis for his $150 million demand. He learned, however, that it was actually paid around $52–57 million. The rule of thumb, as Campanelli knew it, was that the government would usually seek double the amount. He could live going for $7.6 million, but it was not up to him. It was Jeff's and Tammy's decision.

Campanelli had gained confidence in both Peterson and Hastings. He had developed a solid trust with them as much as he had with Jeff and Tammy. Both Peterson and Hastings were also former military, and they shared his outrage. They explained to Campanelli why $7.6 million was reasonable, given the potential for losing the case. Thus he felt if Peterson was comfortable with this amount and if Jeff and Tammy found it acceptable, he could too. If they settled for $5 million, Campanelli thought, it was a practical amount if the DOJ was fair with the relator's share. (The relator's share in a *qui tam* case ranges from 15 to 25 percent of the money returned to the government.)

At their first meeting with Campanelli, Jeff and Tammy explained to him that their motivation was not about the money; it was about doing the right thing. Therefore, they weren't particularly upset with the numbers being floated by Peterson. Campanelli would have liked to see a larger share, of course, but he wanted Peterson to hold firm to the bottom line of $5 million. Peterson told Campanelli that if he was going to try the case, the first thing he would be worried about was the Army coming in, putting on a witness who would testify on the stand in front of the jury that there was nothing wrong with the helmets. In short the Army would not back up the DOJ's case. "Now you are looking at no harm, no foul, and what do you have?" Campanelli reasoned. "You have the false certifications and if you multiply $11,000 times each false certification at a maximum you get $2 million. So if Peterson explains it that way and the bottom line is $5 million, and he believes it to be reasonable, then all right." The DOD and the services siding with their contractors instead of with the taxpayers and the troops cripple many *qui tam* cases.

Campanelli recognized that the government had more information than he had. The federal investigators had conducted a raid and had all the documents. They knew what was going on in the criminal case. He was placing his trust in Peterson. Peterson convinced Campanelli that $7.6 million was a good number based on all the information available to him, information that Campanelli didn't have access to. Peterson told Campanelli that if Sioux Manufacturing didn't agree to $5 million, there was no sense in talking. He was drawing a line in the sand.

The U.S. district court magistrate Judge Karen K. Klein was very successful at settling cases due to her unique ability to bring litigants to closure. Within the U.S. district court in Fargo, it was her pattern to bring all the parties together for settlement negotiations. Peterson had dealt with Klein for thirty years and felt she was one of the best federal judges in the state.

Klein wanted to bring the Sioux Manufacturing case to a close without a trial. She ordered the parties to appear before her on August 22, 2007, to try to settle the case. The parties included the relators, their attorney, and all the defendants with their attorneys. Peterson was very comfortable with Judge Klein being involved in the settlement negotiations and getting it done. However, one issue that was "a little bit dicey" was her preference to have a DOJ representative present during the negotiations. Peterson was doubtful a DOJ representative would agree to appear, but Klein did allow some leeway when Peterson promised that a DOJ representative would be available by phone at any time during the negotiations.

On the second level of the U.S. district courthouse in Fargo is Bankruptcy Courtroom 230, where all the parties would gather for settlement negotiations on the Sioux Manufacturing case. On August 22 Jeff Kenner and Tammy Elshaug drove, in separate cars, the 170 miles from Devils Lake to Fargo—about a two and a half-hour drive. They mentally prepared themselves to face Carl McKay, Hyllis Dauphinais, and the tribal council in negotiations. The case had been going on for a long time, and now they were anxious to get it over with.

Jeff felt sick to his stomach at the thought of going into the courtroom and facing McKay, Dauphinais, and all their attorneys. Tammy shared the same fear. Before going to the courthouse, they walked to the Radisson Hotel, where Campanelli was staying, to have breakfast with him. When they arrived at the hotel restaurant, they spotted Campanelli at a table next to one where both McKay and Dauphinais were sitting. Both Jeff and Tammy backed out of the restaurant and called Campanelli on his cell phone to let him know who was sitting next to him. Campanelli was startled because he had never met them. So instead of meeting in the restaurant, they met in a small room off the lobby to discuss their role in the settlement negotiations. There would not be much for them to do but observe because they would not have any say in the discussions.

Walking into the courtroom later that morning, Jeff and Tammy learned that they would be in the same room with the defendants and their attorneys. Along with Campanelli they sat on one side of the courtroom, while McKay, Dauphinais, their attorneys, and many members of the tribal council filled up the other side. The seating arrangement had the appearance of David versus Goliath. However, nothing happened as they sat waiting for the judge to appear. At first Jeff and Tammy just stared straight ahead. Then Jeff tried joking around—his habit when he was nervous. Tammy was shaking because she had to once again look at the people she had not seen since they fired her. The members of the defense and the tribal officials joked among themselves to pass the time. Jeff and Tammy did not know whose decision it was to put everyone involved together. Concerned about his clients, Campanelli sought out Peterson and Hastings to see if they could be put in a separate room.

Finally, much to their relief, Peterson arranged for Jeff, Tammy, and Campanelli to move to a separate room, a large conference room located behind the courtroom, near the judge's chambers. Because there were too many of them to relocate, the defendants and their attorneys remained in the courtroom. Once Judge Klein arrived, the negotiations began. During discussions Campanelli felt that Judge Klein did not assert much pressure on the parties to settle or offer a solution for how to settle. She shuttled from the plaintiff's side, located in the conference room, to the defense side, in the courtroom, with offers and counteroffers. At the end of the day, Klein announced that Sioux Manufacturing had counter-offered fifty-thousand dollars. The main theme of Sioux Manufacturing's position was that it had done nothing wrong and the case was nothing more than a nuisance. Peterson was angry and let Klein know. Campanelli strongly felt that the counteroffer was insulting. Klein asked Peterson what he was thinking. He told her "seven figures."

Judge Klein left the conference room to consult again with the defendants. She returned with an offer of $100,000, with $50,000 up front and the rest a year later. Peterson, frustrated, thought the offer was ridiculous. "What don't they understand about seven figures?" he said. Campanelli couldn't believe it. "They were offering a payment plan even though they had $12 million in the bank," he said. Klein revealed that Sioux Manufacturing strongly felt there was no case. Everyone in the conference room was upset, and the offer was rejected. The talks continued into the next day. Klein pushed hard on both the relators and the defendants. She kept the defendants' nose to the grindstone, but nothing was agreed on and negotiations ended without resolution.

The settlement discussions came down to the issue of plus-minus thirty-five by thirty-five yarns per inch on weaving. Sioux Manufacturing claimed that rounding up was fine, and it had "authorization" to weave below the thirty-five-by-thirty-five requirement. Its attorneys produced documentation they said was proof of the authorization. According to Campanelli, they produced "garbage." Peterson told him that he had gone through the information, and it said nothing about allowing Sioux Manufacturing to round up or granting a waiver on the thirty-five-by-thirty-five requirement. Campanelli had settled a lot of cases, but this one "was unlike any [he] had settled."

Campanelli was frustrated that he had made a trip to North Dakota for nothing. Sioux Manufacturing's offer "was not a good faith offer," he believed. He sat back and let Peterson take the lead in the negotiations. Jeff and Tammy had been straight with the facts of the case from the beginning. The government had a large amount of evidence even before the raid. Campanelli recalled Shon Hastings telling him about the results of the raid: "We found exactly what we thought we would find." Sitting across from Peterson during the negotiations, Campanelli learned that Peterson had a copy of a letter from a competitor of Sioux Manufacturing, requesting a waiver from the Army to lower the thirty-five-by-thirty-five weave requirement to thirty-four by thirty-four. The Army had denied the waiver, stating that it insisted on the thirty-five-by-thirty-five weave requirement and expected all its manufacturers to carry out the contractual requirements. For the Sioux Manufacturing case, the negotiations were at a standstill.

During the late afternoon of October 15, 2007, about two months after the stalled negotiations, Tammy's husband, Larry Elshaug, drove about ten miles west out of Devils Lake to their private campground at a nearby lake. The purpose of his trip was to pick up his thirty-six-foot camper and bring it back into town to prepare it for the winter. The Elshaugs used the camper about three or four weekends during the summer, usually until Labor Day weekend. While looking for a can of antifreeze he kept in the back of the camper, he found it was empty but wet around the can where the fluid had leaked. He shook the can, and a .22-caliber shell fell out. Larry was mystified about how the shell got into the can until he walked out of the camper and found a bullet hole in its side.

It was dark by the time Larry found the shell, so he towed the camper home, where, upon closer inspection, he found that there were three bullet holes in the side of the vehicle. He then came to the realization that someone had shot at the camper multiple times. One of the bullets had entered under the bunk beds where the kids slept, another entered the dining area, and a third entered up high near the roof. Tammy was immediately concerned that the shooting was inspired by her case against Sioux Manufacturing. The first settlement talks had ended unsuccessfully in late August, so word was out about her and Jeff bringing the lawsuit. Her whole family became nervous. Her kids were scared to go back to the camping area.

The camper was parked on a dead-end road, so access was limited and rarely used. The location is about three-quarters of a mile from Jeff's property on the lake. Larry considered the area normally very safe. The Elshaugs even left their keys in the car out there. They had never heard of any shooting in this remote area.

Tammy's concern led her to call the Ramsey County Sheriff's Department. She was told that she had to come in to be interviewed and file a report. However, due to the court seal on her *qui tam* case, she would be limited in what she could tell the sheriff. She called Jeff, and they agreed that she should tell Campanelli about the incident and ask him how to proceed with the sheriff in his investigation. Tammy phoned Campanelli, and in a measured and calm manner, she told him that someone had fired shots at their camper. She said that everyone in the area recognized her trailer since it was normally parked on her home lot. She thought it might have been neighborhood kids fooling around with a gun but was not sure. Campanelli was immediately concerned, but he felt powerless to protect her or Jeff since he was in New York and they were in North Dakota. "It wasn't a comfortable feeling not being able to protect them," he recalled. He wasn't as concerned about Jeff, who had taken measures to protect himself by keeping a loaded shotgun by his front door and a weapon in his vehicle.

Campanelli informed Peterson and Hastings about the shooting incident. He expressed his concern and explained to them that everyone knew Tammy's camper, it had been parked on a dead-end road, and he felt that the shots were not accidental because it was not in a hunting area. Also the incident was suspicious because it had occurred after the first settlement negotiations, when Tammy and Jeff's involvement in the case was fully revealed to the tribe and Sioux Manufacturing. Campanelli believed the odds were slim that someone would randomly put three shots into the Elshaugs' trailer. "I know we aren't exactly on the same team," he told Peterson, "but do me a favor. To the extent you could do anything to keep an eye on Tammy, please do so. If you could call local authorities, whatever it is, I don't care."

Peterson thought the shooting was unusual given the circumstances and where the camper was located. However, the Elshaugs' camper was not on an Indian reservation. Within two days Campanelli filed an emergency application with the court asking permission to partially lift the seal to disclose only as much information as needed to the sheriff's department to aid in its investigation of the incident.

Judge Klein granted permission, and specific information was provided to the sheriff. In the end the sheriff was unable to resolve the case. It could not be determined whether the shooting was intentionally done to threaten or scare her or to cause her harm or was just random. Tammy nonetheless continued to be concerned for her safety and the safety of her family.

1. Dr. Robert "Doc Bob" Meaders, founder and operator of Operation Helmet, December 9, 2009.

2. Whistleblowers Tammy Elshaug and Jeff Kenner in Devils Lake, North Dakota, February 3, 2008.

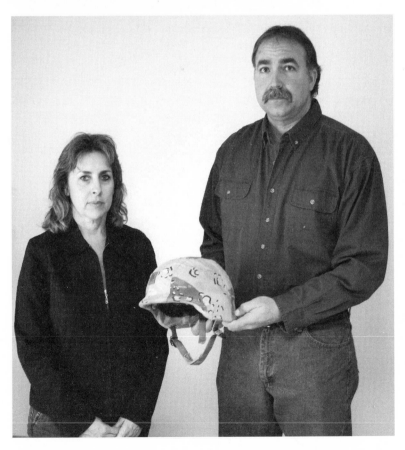

3. View of downtown Devils Lake, North Dakota, February 2008.

4. View of Sioux Manufacturing Corporation, located on the Spirit Lake Indian Reservation, Fort Totten, North Dakota, February 2008.

5. Mike Dennis (right), CEO, and Tony Ericsson (left), COO, of Oregon Aero, Scappoose, Oregon, August 2017.

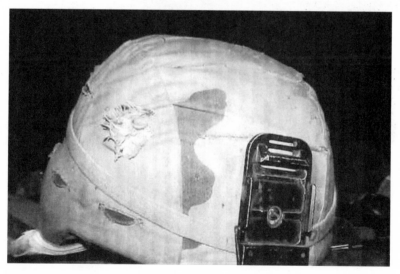

6. Photo of a Modular Integrated Communications Helmet (MICH) typically worn by Special Operations troops in Iraq and Afghanistan, October 2009.

7. Andrew Campanelli, attorney for Jeff Kenner and Tammy Elshaug, in Mineola, New York, February 2009.

8. A view of the Federal Building and U.S. District Court, Fargo, North Dakota, September 2009.

CHAPTER 20

Settlement

In an attempt to kick-start the negotiation process again, Peterson reached out to the defendant's attorneys with a draft settlement agreement using a DOJ template but without a settlement amount included. The details of the draft settlement didn't go over well with the attorneys. Not long afterward, Peterson received their version of a settlement draft that was almost completely unacceptable to him.[1]

By now both Peterson and Hastings were becoming completely frustrated by the defendant's response, which is evident in Hastings's email to Campanelli: "You will be offended by SMC's redraft," she wrote. But she quickly advised him, "Before you spend a lot of time re-writing, please know that virtually none of the redraft will be accepted."

However, what really unnerved Peterson was a call he received from Sioux Manufacturing's counsel announcing that UNICOR had just signed another $14 million contract with Sioux Manufacturing to weave Kevlar pattern sets for the ACH. Peterson was offended. It wasn't what he wanted to hear from the opposition, and he was outraged at the announcement, right in the middle of settlement negotiations. What also disturbed him was that he didn't hear it from UNICOR first. He had a good working relationship with the UNICOR attorneys, but they did not give him a heads-up on what they were going to do. Why couldn't they have waited to announce the contract, he wondered, or at least tell him in advance? Additionally, Peterson learned that somebody from UNICOR had given Carl McKay a golden helmet award in recognition for outstanding service to the military. Peterson nearly came unglued when he heard that.

As far as Army procurement officials were concerned, a fraud case against a contractor had almost no effect on their procurement decisions or the timeline of announcement. If there was no suspension or debarment action against a contractor, they would continue to use the contractor despite fraud allegations or a pending settlement. So a contractor could continue to hide improper behavior or stall a case while procuring a follow-on contract that allowed the contractor to continue acting in bad faith. It is still how the Pentagon often does business with its favored contractors.

With the sudden announcement that UNICOR had awarded another contract to the very people who had perpetrated the underweaving, Peterson believed it caused the dollar value of the settlement to plummet. It decreased with each revelation of this sort. First, Peterson emailed Campanelli, advising him, "We just found out that UNICOR has apparently entered into a $14,000,000 contract with SMC to provide Kevlar pattern sets for the new ACH helmets. We found that out from SMC lawyers and not UNICOR. I expect you can appreciate my reaction to that news and that you can also appreciate the wonderful cross examination possibilities that provides for SMC's lawyers in the event of litigation, and the effect such activity has on settlement negotiations." Hastings followed with an email to Campanelli that the settlement amount would now be between $1.5 million and $3 million. Campanelli wrote back, "Speaking quite candidly, I do not know if that they [relators] will agree to a settlement below $2 million, irrespective of what advice I may give them. While I find it hard to fault them for it, as a matter of principle, Jeff and Tammy view the defendants as 'getting out of this' for less than what they saved in Kevlar."

Campanelli could sense the wind being taken out of Peterson's sails with the contract award revelation. For him it was a smack in the face, a move that was wrong on many levels He believed it was the tipping point when negotiations really started going south. Peterson could sense Campanelli's frustration and the possibility that the relators might not agree to a lower settlement amount. He emailed Campanelli to let him know, in the strongest terms possible, that the DOJ might not intervene in the case if the relators do not agree to settle: "I also understand that you are troubled but I think you and Jeff and Tammy need to understand that IF we do not get this matter settled NOW, the likelihood is that the U.S. may not intervene and you will end up litigating the case."

If the DOJ did not formally join the case, it would be up to Campanelli with Jeff and Tammy to take the case through the courts themselves—an expensive and daunting task for a complicated case. Peterson added that despite how prepared Campanelli might be to take the case on, Sioux Manufacturing would "NEVER SETTLE and [would] fight [him] and the relators to the end."

Campanelli believed that either UNICOR or the DOD was sending a message with the timing of the contract award and saying that they were keeping Sioux Manufacturing as a contractor despite threats that the case could result in suspension or debarment. He also believed that given this message, the DOJ was now applying pressure on both Peterson and Hastings to lower the settlement amount and make the case go away. Campanelli had a candid conversation with Jeff and Tammy about all the maneuverings, which upset them both.

Jeff Kenner was so disgusted by the way settlement negotiations were going that it constantly flooded his mind. He felt they were being put in a position of having to settle low, and in the end Sioux Manufacturing would be vindicated. In the middle of the night, he woke up after dreaming of words to express his feelings about the case. He quickly got out of bed to write down what he had dreamed. "It was like you had to throw up and couldn't stop it from happening," he recalled. Writing down the words in the quiet of his home without distractions, he made them into a poem:

And now what do I think across that I have stumbled.
To tell and tell, truth I have, upon deaf ears it fell.
The good side used all I had and it seems they only fumbled.
I can only imagine now what the evil side will likely fake.
Retaliation with words and deeds I pray that they won't make.
Now lies from their devious minds have come to the light.
It's up to me to win with truth, evidence and might.

As settlement negotiations continued outside court between Peterson and the defendant's attorneys with no agreement on the horizon,

Judge Klein decided to bring everyone back together for another try. She called for a settlement conference for November 7, 2007, in Fargo. Peterson believed Carl McKay, through his attorney, was making overtures to the judge that might have gotten them back to the table a second time. The difficulty Klein was having with the tribal council was its unwillingness to put up the necessary amount of money.

In addition to the stalemate in settlement negotiations, the defendant's attorneys began questioning why the relators should get a percentage of the money from the settlement as required under the *qui tam* law. Campanelli felt it was odd that the defendant's attorneys didn't know that by law relators received a percentage of money obtained by the government. In order to "educate" them, he sent an email to Larry Leventhal, the attorney for the tribal council, on November 3—four days before the settlement conference—saying, "Your letter reflects that you do not possess even a basic understanding of the False Claims Act (FCA) as it applies to the Relators in this case." Further, he explained that a relator's share is statutorily required under the FCA. Also the FCA provides that the relators shall receive reasonable expenses the court finds have been incurred, plus reasonable attorney's fees and costs. It was Campanelli's impression that Vogel and Leventhal were "absolutely clueless" about how the FCA worked. Hastings was not amused at Campanelli's email. She emailed him, underscoring her concern by sarcastically saying, "I'm sure your tone will help move things along at the settlement conference."

On Wednesday, November 7, the second round of settlement negotiations was set to begin in Fargo. The day before, the defendants filed a sixty-four-page answer to the complaint. In general the answer denied all charges that Sioux Manufacturing had woven Kevlar cloth under thirty-five by thirty-five yarns per inch. In fact it claimed, "Between 1987 and the present time, SMC consistently manufactured 35 x 35, 2 by 2 basket weave, Class II, Type 2 fabrics, all as required by the contracts, the military specifications and MPS-KH-0333 (SMC's Manufacturing Process Specification)."

However, in the answer, the weave below thirty-five by thirty-five was justified by claiming that the military specification for Kevlar cloth allowed thirty-four-by-thirty-three and thirty-four-by-thirty-four weave density prior to 1987, and the ballistics requirements did not change after the specification was changed to thirty-five by thirty-five. "Thus it does not appear that the ballistic protection that is required to be provided by a 35 x 35 thread per inch weave is jeopardized or that helmets woven with lower count thread count cloth, such as a 34 x 33 or 35 x 34 weave cloth, would fail ballistics tests or be unsafe or a threat to human life," the answer contended.

In addition to denying that the company wove Kevlar fabric of less than thirty-five by thirty-five density, Sioux Manufacturing claimed in its answer that all manufacturing personnel were instructed to follow the company's internal specification that required a thirty-five by thirty-five weave. However, Campanelli's response was that employees were not instructed to follow specifications as evidenced by the fact that the weaving supervisor, who had been with the company over twenty years, didn't know what the specifications were. Also the quality assurance manager apparently didn't know what the specifications were as evidenced by a company document that described the specifications as thirty-five by thirty-five, plus or minus one.

The answer acknowledged the termination of Jeff and Tammy in November 2005 but denied that it was the result of their revealing the underweaving issue. According to the answer, the only reason Sioux Manufacturing had terminated them was that "they had violated company policy." Using the affair as a reason, it further claimed, "Although both were married, they appeared to have an inappropriate relationship with each other, rather than a business relationship. This conduct was tolerated, although they were advised that this conduct was not proper." Sioux Manufacturing was determined to stick to the old affair canard despite a complete lack of evidence or proof. Also it denied that removing Jeff as a weaving supervisor was in response to allegations of underweaving. Instead, it explained that Jeff was only a "temporary supervisor," so there was no demotion.

Actually, the major problem with the PASGT helmet, according to Sioux Manufacturing, was not the Kevlar but its sling suspension system. It did not provide adequate protection from injuries caused by blast because it lacked shock-absorbing pads, which were installed in the new ACH, as Dr. Bob Meaders had testified before the House

Armed Services Committee Subcommittee on Tactical Air and Land Forces in June 2006. The company asserted it had nothing to do with producing the suspension liners in the PASGT nor was it involved in the decision not to improve the suspension system in the PASGT. It was an ironic moment when the company that was willing to skimp on the safety of its part of the helmet manufacture blamed the flawed inside design of the helmet and used Doc Bob's testimony to cover its wrongdoing. Sioux Manufacturing's rebuttal had inadvertently merged these two scandals.

The settlement negotiations resumed in the same courtroom of the federal courthouse. Because UNICOR had recently awarded Sioux Manufacturing a new contract, Peterson started the negotiations much lower, at $2.2 million. But Sioux Manufacturing was still intransigent and refused the offer. It was still stuck at $100,000. At that point Campanelli didn't think Sioux Manufacturing would come up to $2 million. He was pretty much resigned to becoming the litigator on the case when the DOJ dropped out.

Located across the street from the courthouse sits the Plains Art Museum, a large, three-story tan brick building. On the first floor of the museum, in a large open-beam ceiling area is the Café Muse, where many court and federal building personnel and others attending court proceedings go for lunch. During a lunch break in the negotiations, Jeff, Tammy, and Campanelli met for lunch at the café. Soon after they sat down to eat, both Peterson and Hastings walked in and sat across the room from them. When Peterson spotted them, he went over to their table and said he had just got a call from Washington. The DOJ's position was that if the case didn't settle for $2 million, the government would decline to intervene in the case.

Campanelli believed that by then he had developed a pretty good relationship with Peterson and Hastings. Peterson was telling him what Washington was going to do if the case was not settled. He had to view it with the backdrop of the email Peterson had sent to him, predicting that the defendant's attorneys would bury him if the DOJ didn't intervene. The defense attorneys would fight Campanelli tooth and nail to the end if for no other reason than to make money from their high fees. Campanelli's impression was that the decision to decline would not be based on the strength of the case because they had the evidence needed to sustain the case. Peterson then revealed that if the case went to trial, a witness from the DOD would take the stand and say that all the helmets passed ballistic tests.

Once back in the courthouse, the judge spent about an hour with the defendants while the relators, Campanelli, Peterson, and Hastings waited patiently, telling a lot of jokes among themselves. To Jeff it felt like Sioux Manufacturing was running the show and pulling strings on much of it.

After more give-and-take in the negotiations, both sides finally settled on $2 million. The defendants agreed to pay $1,935,000 as the settlement amount. Carl McKay would later deny any wrongdoing, but as many companies do, Sioux Manufacturing justified agreeing to the settlement on the grounds that the legal costs would have exceeded the settlement amount if it had continued the fight.

As much as he hated the low offer, Campanelli recommended to Jeff and Tammy that they take the settlement. The money was on the table, and if they went forward and lost, it would end up being zero. He let them know that he would have trouble sleeping at night as their attorney because of all the time, risk, and suffering that they had experienced. Campanelli believed that Sioux Manufacturing was now weaving the right way, and thus they had actually won that part of the case. Jeff was upset with the settlement because he felt that Sioux Manufacturing got off with only $2 million after putting the troops at risk.

It's not unusual for *qui tam* relators to believe the recovery amount to be small relative to their perception of the extent of the fraud and the disruption and damage to their careers and reputations. Yet despite their dissatisfaction with the settlement, Jeff and Tammy still felt the *qui tam* action was important and needed to be pursued.

After the agreement was reached between the parties, it was finalized on November 8. Jeff signed it reluctantly because he thought more could have been done. Written within the agreement, the relators and the United States alleged that Sioux Manufacturing "did not always comply with the 35 x 35 weave density standard"; however, the company continued to claim that it fully complied with all military specifications and "dispute[d] all of the material allegations of the Relator's Complaint." Thus the agreement was "neither an admission of liability by defendants nor a concession by the United States or the Relators that their claims [were] not well founded." The contentious case took until December 10 for all parties to sign the agreement.

With the agreement signed and a judicial order to unseal the case issued, the next battle was over the DOJ press release. Versions bounced around between the U.S. attorney's office, the DOJ, Sioux Manufacturing, and the tribe. Some fifteen different versions were drafted, with almost as much negotiation over that language as there had been over the settlement itself. Eventually, the DOJ decided on a final version. Peterson felt it was a watered-down version of a watered-down press release. In the end the DOJ and Sioux Manufacturing each issued a separate version of the press release.

The DOJ press release revealed that the allegations "arose from a lawsuit filed in the District Court of North Dakota by two former employees of SMC." The Sioux Manufacturing press release claimed that it "came as a result of an allegation from disgruntled former employees." DOJ's release stated, "The investigation found evidence that, on occasion, SMC knowingly delivered cloth that had not been woven to the precise specifications." Sioux Manufacturing countered, "We deny any and all of the allegations originally brought to the attention of the DOJ by disgruntled former employees. . . . It is unfortunate such allegations can be made that don't require the burden of proof." It added, "Not once during the investigation has the DOJ ever shown us one scrap of non-compliant fabric, or made reference to possession of such."

The case ended with no helmet recall, continuing Army resistance to cooperation, no admission of guilt, and no guarantee that Sioux Manufacturing would not slip the weave count in the future. The main deterrent was the negative image for Sioux Manufacturing, but the Army made sure that the contracts kept flowing with little consequence. Tammy and Jeff had to live with a moral victory, knowing that they had done the right thing, even though only a small part of the fraudulently gained money was returned to the federal government.

CHAPTER 21

Just Give Me What I Want

For twelve years the Military Free Fall School located in Yuma, Arizona, used Oregon Aero pads for its free-fall helmets even after Oregon Aero's disqualification by the Army and Marine Corps helmet bureaucracies as a vendor for the ACH and the LWH. The school trains many of the elite Special Forces operatives from the Navy SEALs, the Marine Force Reconnaissance, the Air Force Para-Rescue, and the Army Rangers in the parachute techniques of high altitude-low opening (known as HALO) and high altitude-high opening (known as HAHO).

It didn't matter to Mr. Helmet, George Schultheiss, that the Free Fall School was satisfied with Oregon Aero's pads. Again pushing his green foam as he did in 2001 (see chapter 8), he wanted Oregon Aero to purchase his own green-foam pads for the free-fall helmets. On February 22, 2007, Schultheiss emailed Oregon Aero's Tony Erickson, claiming, again, that he was the original designer of the "Protec MFF Liner when it was transitioned to Oregon Aero back in the mid to late 1990's." He said, "I am recommending to this user at Yuma [Free Fall School] to only use MFF liner pads with the Green Confor foam." Schultheiss directed Oregon Aero to provide pads "utilizing only the Green Confor foam from EAR [Confor foam is a proprietary urethane foam made by E-A-R Specialty Composites]." Also, he asked Oregon Aero to provide test data "supporting the blue/pink pads over the green pads especially if the data show[ed] better impact protection in all temp ranges."[1]

Erickson replied in an email that Oregon Aero had always used their blue and pink foam for HALO operations and had never produced strictly green foam pads. Also, he pointed out that "early development with the ballistic helmet pads indicated superior impact protection in the current configuration versus configurations that utilized the green only foam."[2]

In a follow-up telephone conversation with Tony Erickson and sales representative Lee Owen, Schultheiss suggested they continue to sell pads to the free-fall school but change their pink and blue foam to his green foam. However, Owen made it clear to Schultheiss that they had no interest in changing to another type of foam, especially since they had already invested in the pink and blue foam. In earlier testing Oregon Aero found that the green foam was less user-friendly, less conforming, and was stiff and unforgiving. When placed in helmets, it would tend to pop loose and fall out of the helmet.

Owen asked Schultheiss if he had any data that would compel Oregon Aero to change to his green foam. Ducking the question, he responded that he was more interested in seeing what data Oregon Aero had, and he didn't have to show them his data. Instead, with no evidence that any test data existed for the green foam, Schultheiss referred to some "historical data" but claimed that he didn't have access to it. He explained that it was too old and may not have been retained. However, he did have access to Oregon Aero data for the MICH and the ACH. Schultheiss recalled testing samples ten to twelve years earlier with the surgeon general at Fort Rucker with other foams and found that Natick had the most protective and user-acceptable product with the green foam.

Owen reminded Schultheiss that Oregon Aero's foam pads had been used for free-fall helmets for twelve years with no change to the product. He asked Schultheiss why a change was needed and why there was such a compelling, unalterable reason to use his green foam. Schultheiss claimed that his foam offered better protection then the pink and blue foam. He admitted that Oregon Aero's pink and blue foam was the right item for the MICH, which weighs much more than the free-fall helmet. Further, he contended that because the flexible free-fall helmet is a lighter helmet, the conditions were different with the military free-fall helmet, although the requirements were the same. Schultheiss claimed that he was seeking to maximize protection as much as possible.

Owen then asked Schultheiss if he was willing to fund a comparison test between Oregon Aero's pink and blue foam and his green foam. Schultheiss rejected the offer, claiming a lack of funding. Claiming no problems with their product in twelve years of use by the free-fall school, Owen again asked why he wanted to change to his green foam. After a long silence, Schultheiss said it was his mission to have the best protection possible for the operators. It was their mission also, replied Owen. Still, Schultheiss believed that Oregon Aero's foam did not offer the best protection.

Schultheiss continued to press his demand, claiming his green foam was installed in a Protec helmet that sat in his office, and it worked. (Protec helmets are half helmets used primarily for recreational purposes by skaters, skiers, kayakers, and bike riders.) Owen suggested that he send the Protec with his green foam, and Oregon Aero would assess it. Schultheiss declined the invitation, claiming that he had to discuss it with other people first.

Trying to let Owen and Erickson know that he was the one in charge of the Free-Fall School procurement, Schultheiss chastised them for refusing to agree to use his foam when he asked them to do so. He expressed puzzlement at Oregon Aero's refusal to make his green foam and sell it to the Free Fall School. He blatantly told them that the government did not have to prove that his requirement was what he wanted or not wanted. "Just give me what I want," he demanded, according to Erickson. "Give me a quote on the green foam. You are saying no, you should say yes."[3]

Despite Schultheiss's reluctance to fund a comparison test, Oregon Aero decided to conduct one anyway. Schultheiss, in another conversation, told Erickson and Owen that if they tested the pink and blue foam against his green foam and the pink and blue worked better than the green, he would accept the results. Oregon Aero took him up on this offer and conducted the comparison test using an independent testing lab to avoid bias. The results indicated that the Oregon Aero pink and blue foam afforded about 60 percent greater blunt-impact protection than the Schultheiss green foam.

Furious at this result, Schultheiss disparaged the testing and accused Oregon Aero of not conducting the test correctly. Trying to assert his buying power, Schultheiss said that he was going to shut down the Special Forces HALO helmet because he made the decisions on the pads and he didn't think it would be a safe helmet with Oregon Aero pads, never mind that the HALO helmet had used Oregon Aero pads for twelve years without safety problems.[4] However, nothing more was heard from Schultheiss on the matter. In the end Oregon Aero continued to supply its pads to the Free Fall School.

It's difficult to determine the rationale for Schultheiss's actions, whether he was just trying to get credit for creating a foam pad or trying to force Oregon Aero completely out of all military sales. Military bureaucratic logic can be hard to discern—was it to bridge the differences between the bureaucrat, who is wedded to his official actions or his private motives, and the vendor, who is wedded to its profit motive? The situation often resembles two people talking past each other. It was not the first time that Schultheiss had tried to push his green foam on Oregon Aero without success. He was upset with Oregon Aero patenting its product and claimed that it had stolen his idea for the green foam even though the products were not alike. His actions were typical of a bureaucratic "czar" according to defense analyst Pierre Sprey. "They want to be the chief design engineer," he explained. "He's telling them what shape it has to be, how thick it has to be, and what pads to put in it. They always try to exercise their chief designer authority."[5]

Oregon Aero had to fight against Schultheiss's bureaucratic logic and touchy ego. Since 2004 the company had been ousted as a pad vendor for the ACH and the LWH due primarily to bureaucratic decisions based on faulty, nonexistent, or possibly rigged tests, along with higher cost. By 2007 it had tried meetings and other types of communications with Natick, USAARL, and PEO Soldier to resolve those issues, all without success. It didn't seem to matter that by early 2007, Operation Helmet had sent more than thirty-five thousand Oregon Aero pad kits to the troops who requested them over the Army and Marine Corps supply-issued pad systems.

George Schultheiss eventually moved from the Army's helmet program at Natick and became an engineer at Natick's Research, Development and Engineering Center. He died unexpectedly in February

2017. However, the pad procurement problem was not limited to Schultheiss. He was just one individual in a bureaucracy more concerned with itself and its funding and survival than with its mission to properly equip the troops.

During the early part of 2007, Tony Erickson met with the National Institute for the Blind in the hope of moving forward to vendor qualification. The staff of Oregon Aero had a year to adjust to the reality of its situation with the Army, and they were now somewhat jaded. The president of the NIB, a tall, angular, and stark man, presided over the meeting. He stopped the presentation about a third of the way through, stood up, and said to the Erickson that the NIB knew all about Oregon Aero and saw the company as competition that took business from it. The NIB made the webbing for the padless, sling-suspension system used in the PASGT helmet and early LWHS. However, the NIB president admitted the institute also knew that Oregon Aero produced the best foam pads that everyone preferred.

He then revealed that the problem with Oregon Aero was not a qualification issue but rather that the company was too expensive. He claimed that the Defense Logistics Agency set the price, and NIB couldn't purchase the pad kits for more than that price. Oregon Aero's price of around fifty dollars was too high for the NIB, and Team Wendy's product was half that cost. He pleaded with Oregon Aero to help the NIB buy its product because he knew it was the better pad system. He admitted that the NIB had seen the Operation Helmet website and the emails from soldiers who were unhappy with the pads that came with the ACH because they did not work as well as Oregon Aero's pads.

The NIB president's bottom line was that Oregon Aero should help the NIB sell its pads to the DLA so that agency would agree to pay more for Oregon Aero pads than for the other manufacturers' pads. Erickson believed that for a protective device to have the tail wagging the dog was backward. It shouldn't be argued that given the proven merits of Oregon Aero's product, the decision should be based on best value. Instead, price governed the decision irrespective of quality or how the product functions in combat.[6]

Paul Diamonti, director of product development fo r the NIB, stated in its 2007 request for quotation for helmet pads that the NIB's primary goal was to provide the military with the best-performing pad systems at a fair market price while maximizing the possible amount of direct labor operations for the blind employees working in its associated agencies. Reporter Art Levine believed that it was important to recognize that the NIB's goal contained three competing requirements: achieving maximum employment for the blind, acquiring the best-performing helmet pads, and providing helmet pads at a fair market price.

Levine argued that the inescapable reality was that the NIB decision maker can only choose the best one of those three goals, not all three simultaneously. He felt that the NIB had elected to purchase helmet pads that were inferior to the Oregon Aero design. This decision, he believed, significantly contributed to the TBI problem and led to conclusions that the NIB was not technically qualified to evaluate helmet-pad performance.

The NIB was also not capable of comparing a vendor's price to those of its competitors in the same procurement or to costs realized in similar, recent procurements. Most importantly, according to Levine, the NIB placed the greatest emphasis on maximizing the amount of direct labor operations for blind employees at the expense of American soldiers. While the NIB was to be applauded for its service to blind and vision-impaired people, Levine concluded that the facts presented above illustrated that the NIB was neither competent nor equipped to be entrusted with procurement decisions affecting the safety of American troops.

Oregon Aero's two years of queries to determine who actually was responsible for purchasing the pads for the Army led to no solid answers. The company could not discover who within the Byzantine labyrinth of military procurement was actually responsible. Bureaucratic layering and muddled communication between and within military agencies made accountability impossible to determine. Military procurement, when executed by the DOD's complex, confusing, acronym-laden, intertwined agencies, is the most complicated, confusing, and inefficient procurement system in government. Very few people truly understand it. Its complexity and lack of transparency make it vulnerable to fraud, waste, abuse, and especially manipulation for career ambition. Understanding military procurement makes solving stochastic calculus problems seem like basic first-grade math.

While trying to determine who made the procurement decisions for helmet pads, Oregon Aero was told by Natick that the NIB had the contract and chose the pads to be purchased. However, the NIB said it didn't have testing or evaluation facilities and had no way of evaluating the pads, thus it did not make the procurement decisions. It claimed that the Defense Logistics Agency made the procurement choice, and the NIB only assembled the pad kits. Undeterred, the DLA passed the buck back to Natick, claiming that Natick told it which pads were approved and which ones it was to buy. Natick's helmet officials claimed they didn't make the decision—it was their "boss," a mid-level administrator at Natick who had that responsibility. The "boss" denied responsibility for the decision-making, claiming his engineering department at Natick made that decision. However, Natick engineers deflected accountability, asserting they didn't have that responsibility and only made recommendations. This dysfunctional system was a classic example of the dog chasing its tail but never catching it.

To Oregon Aero the reality seemed to be that one person in one key position manipulated the whole show. The ringleader could be Schultheiss or someone like him dictating which pads were to be purchased. It was someone entrenched in the bureaucracy who controlled the whole machine and whose decisions were greased up the chain of command. Who, then, tells the DLA or the NIB what to buy? Oregon Aero's thought was to flush the whole system down the toilet and not try to fix it. The bottom line was that nobody in the Army took ownership of the decision for pad procurement.[7]

CHAPTER 22

Kafkaesque Nightmare

During the early days of the war in Iraq, Brig. Gen. Mike Caldwell, commanding officer of the Oregon National Guard, received a phone call at his office in Salem, Oregon, from one of his executive officers with a unit of Oregon guardsmen deployed in Iraq. The executive officer was excited about a new padded suspension system Caldwell had sent to the unit for use in its PASGT helmets. "My God, where did you get these things?" the executive officer asked Caldwell. "They are the greatest things since sliced bread." Caldwell had previously been visited by a couple of Oregon Aero officials, who introduced the pads to him. He thought they were amazing, a great fit, and he was especially impressed with the comfort. It made the helmet work a lot better. This was also Caldwell's introduction to the padded suspension system. He was sold on its use for his guardsmen who were already deployed in Iraq or soon to be deployed.[1]

Caldwell joined the National Guard in 1971 as a private. His father had wanted him to join instead of being drafted. Caldwell didn't mind being drafted, but his father was not so pleased. As a World War II vet, he thought it was a big mistake if his son did not join the Guard and go to war with people he knew. Caldwell stayed in the Guard but would also be elected to the city council in LaGrande, Oregon. Three years later he was elected county commissioner. During this time he was also rising through the ranks of the Guard. In 1988 he took a full-time job with the Oregon Military Department as the public affairs officer, a job that included a lot of political work. His National Guard work in that position resulted in a promotion by the state to brigadier general, a position he had filled full time for the last twenty-two years.

Caldwell was completely satisfied with Oregon Aero's padded suspension system, especially with the feedback he was getting from his units in Iraq. Thus in 2004 Caldwell decided to pursue a congressional appropriation to purchase more Oregon Aero pads for their PASGT helmets. Tied in with the Florida National Guard, the Oregon National Guard received an appropriation of about $15,000, which enabled it to purchase the pads. According to Caldwell the Guard received approximately two thousand to three thousand Oregon Aero pads with that appropriation. Caldwell spread the word to other Guard organizations that pads were available for Guard units deployed in the Middle East battlefields.

Not long afterward the Army transitioned to the ACH, which included a supply-issued padded suspension system, one that Caldwell felt was not as good as the Oregon Aero pads. He noted the difference between two of the pads used for the ACH, those from Team Wendy and from Skydex. He pointed out "one of the pads was like packing material you would find in a cheap bicycle helmet while the other one was as hard as a rock"—this "versus something the soldiers could literally sleep with their helmets on because it was so comfortable."

Once the Army fielded the new ACH to the Guard, the Oregon National Guard soldiers, used to the Oregon Aero pads for their previous PASGT helmet, started questioning the supply-issued pads furnished with the ACH. They were disappointed that they did not get the ACH with Oregon Aero pads and wanted to replace the supply-issued pads. Unfortunately, the soldiers had turned in their PASGT helmets along with the Oregon Aero pads when they were issued the ACH. Caldwell started hearing about soldiers buying Oregon Aero pads out of their own pockets or obtaining them from Operation Helmet to replace the supply-issued pads.

One of the Oregon guardsman in Iraq wrote Doc Bob after receiving Oregon Aero kits from Operation Helmet:

Most guys were upset when they transitioned to the new ACH and had to give up the old K-pots with the Oregon Aero pads. As for the problems we've had with the ACH pads, they don't seem to be built for much comfort at all, many soldiers complain of headaches and a general uncomfortableness after just a little while when on missions wearing the issued pads. They seem to fall apart rather fast after awhile, this is due to the wear and tear from sweating and dirt, then being washed, and the cycle being done over and over again. Those are the two biggest issues we've had so far, the pads are not comfortable for long wear, they're hot and they tend to deteriorate fast from wear and tear. The Oregon Aero pads seem to offer the same protection but are more comfortable over long periods of time. When you're wearing the ACH for upwards of 6–10 hours straight comfort means a lot.

To Caldwell there was no question that his troops wanted Oregon Aero pads for their ACHS. To him it was necessary "in order to provide safe and comfortable padding for his soldiers that [would] be in harm's way when deployed to Iraq." At the same time, the Army admitted a serious problem existed with soldiers sustaining TBI. Up until then it had been something the Army was reluctant to admit, but now it had become a major issue. Caldwell started seeing TBIS once his units began arriving back from Iraq in 2006. He knew of the MICH—it was in the Special Forces system at the time. He was also aware that Oregon Aero pads were used in the MICH. The Army was providing Oregon Aero pads to its Special Forces units, and Caldwell wanted his Guard troops to get what the Special Forces had.

Caldwell and his procurement personnel immediately pursued Oregon Aero pads through the National Guard Bureau. They learned, however, that their soldiers at the start of their deployment obtained their gear from a National Guard central issue facility, where they received whatever gear the facility had in stock. It was supposed to be the latest and greatest, but the Army and the National Guard Bureau were buying the cheapest pads they could get, which cost about fourteen or fifteen dollars a set while the Oregon Aero pads were running around forty or fifty dollars a set. Caldwell argued that the comfort issue and protection provided by Oregon Aero pads were better and that the Army should spend the extra money because the soldiers wanted the comfort and deserved the protection.

In 2007 Caldwell made his next move with the assistance of the

Oregon congressional delegation to pursue another earmark as part of the fiscal year 2008 Defense Authorization Act for the Oregon Guard to purchase Oregon Aero pads. Unfortunately, this effort soon resembled a Kafkaesque nightmare.

During Caldwell's effort to obtain the assistance of the Oregon congressional delegation, he received a call from an Oregon congressional representative's office offering its help to obtain Oregon Aero pads. Caldwell said the Oregon National Guard would love to have the pads in all its helmets. The soldiers liked them, and the Oregon Guard would support the congressman to seek an earmark. On March 16, 2007, Representatives David Wu, Darlene Hooley, Earl Blumenauer, and Peter DeFazio officially made a request to the chairman of the House Appropriations Committee to fund Oregon Aero ballistic helmet liner kits as an earmark in the fiscal year 2008 Defense Appropriations Bill.

The support of the Oregon congressional delegation succeeded in obtaining a $1 million earmark in the fiscal year 2008 Defense Appropriations Act. Caldwell soon was shocked, however, to learn that most of the earmark money called for Skydex and Team Wendy pads. Then he noted that congressmen from Ohio and Colorado, where those two companies were located, had also made requests on the same earmark. Local politics, rather than need, had infected the entire earmark.

All combat gear for the Oregon National Guard is stocked at the Portland, Oregon, U.S. Property Fiscal Officer, Central Issue Facility (USPFO-CIF). The facility handles the money, made the purchases, and technically reported to the national Arlington, Virginia, National Guard Bureau, not the Guard adjutant general of each state. Each state has its own USPFO located in it. This system of checks and balances is designed to ensure that purchasing follows DOD procurement rules.

The National Guard works for the governor under U.S. Code, Title 32, the National Guard statutory language. Entities that work for the Army are covered under U.S. Code, Title 10. The USPFO in Portland is a Title 10 officer. He works for DOD but is housed with the Oregon National Guard and is a regular guardsman. His paycheck comes from the chief of the National Guard Bureau. Although 3 percent of the Oregon National Guard's expenditures comes from the state and is used for administration and facilities, most of the Guard's equipment is purchased with federal money. The Guard could have purchased the pads with state funding, but that source was not available. As Caldwell learned, this convoluted system, which was supposed to provide checks and balances, instead would succumb to the political helmet-pad wars.

The Guard Material Management Center (GMMC) is run by the Kentucky National Guard in Lexington, Kentucky. Located at the center is a National Guard Bureau facility, staffed by personnel from the Kentucky National Guard. Lexington is also the location of a large Army depot that is used to move materials and equipment in for stock and out to soldiers. Congressional appropriations for the National Guard become a line item in the National Guard Reserve Equipment Account (NGREA). The Guard obtains its uniforms, boots, helmets, and so on through appropriations earmarked for the NGREA and then distributes them to Guard units from the CIF in Kentucky. This system, which has existed some fifteen to twenty years, was designed to keep the Guard in up-to-date equipment as opposed to hand-me-downs, and it has usually worked well.

It was the job of the Portland USPFO-CIF to purchase the pads for the Oregon Guard with the earmark money. Under the acquisition process, the Defense Supply Center in Philadelphia is a part of the approval process. Because the purchase exceeded $100,000, the approval also had to go through the Oregon National Guard legal office (JAG), and because it could potentially exceed $550,000, another approval had to come from the PARC (principal assistant responsible for contracting) located at the National Guard Bureau.

After receiving the earmarked funds in early February 2008, USPFO-CIF Portland officials faced a multitude of obstacles to acquire Oregon Aero pads. First, they learned they could not make a sole-source selection for the pads. Then they discovered that Oregon Aero was no longer an approved vendor for main Army purchases. The pads they ordered would be Team Wendy pads obtained from the NIB. The CIF-Portland was surprised because Oregon Aero had an approved pad system for use in the PASGT helmet, and the pad system was well liked by the troops. At the time 5,000 PASGTs and 1,400 ACHs were in use by the Oregon National Guard.

There are fifty-four USPFO facilities throughout the country, and some of them specialize in certain equipment, making their personnel the go-to sources for those specific items. The USPFO-Arkansas facility specialized in helmet pads, so the CIF-Portland made an inquiry to the Arkansas USPFO, asking why it could not get Oregon Aero pads. The Arkansas officer's response cited the infamous 2005-12 USAARL test report as the reason that Team Wendy helmet pads had been chosen by the Army as the standard-issue pad for the ACH. Skydex and Oregon Aero pads were not approved.

Caldwell would not be dissuaded. He continued to try to purchase the Oregon Aero pads that the earmark targeted. Regardless of the political problems and other issues, he believed it was the right thing to do for his soldiers. It wasn't long after the earmark was approved that Caldwell received a phone call out of the blue from an Army lieutenant colonel. Although Caldwell couldn't recall the officer's name, he believed that he was associated with the Army's PEO Soldier and responsible for procurement and fielding of soldier equipment, including helmet pads. The lieutenant colonel tried to explain to Caldwell why he could not get Oregon Aero pads. He told Caldwell, "The problem is not so much the comfort issue; it's which one of the pads gives the most protection and Team Wendy and Skydex exceeded the requirement where Oregon Aero has failed." The lieutenant colonel was very candid, telling Caldwell that he wasn't certain the tests were done adequately and that Oregon Aero had basically been "screwed in the testing."

The lieutenant colonel implied that Mike Dennis, of Oregon Aero, had also upset Army helmet officials, those who made the procurement decisions on pad vendors, which had "poisoned the well," and they had personally vented their anger on him. He further explained that there was basically only one approved pad system: Team Wendy. Caldwell had been unaware that Oregon Aero pads had failed the tests, but the lieutenant colonel confided to him that a second round of testing was going to be done because of problems with the first round. Oregon Aero would be invited back to test its pads.

Caldwell told the lieutenant colonel that he didn't know Mike Dennis very well, but he had been around him long enough to know that it wouldn't take a lot for him to raise some people's ire. Dennis is a brilliant guy who comes off as the smartest guy in the room, Caldwell explained. He was very self-assured and very confident about his product and the research and science behind it. It wouldn't take much for a guy like him to "piss someone off." The lieutenant colonel remarked, "You are very perceptive."

Undeterred, the Oregon USPFO tried to obtain an acquisition waiver through the Defense Supply Center, Philadelphia, but it would not grant the waiver. At the same time, the U.S. Army Acquisition Service Center (USAASC) reported that the only authorized ACH pads were those provided by the NIB (Team Wendy pads) or the MSA (Mine Services Appliances). The bottom line was that the Oregon National Guard could not buy the pads it wanted, and Caldwell was stuck in a circular bureaucratic mess.

Caldwell had not heard of any restrictions of Oregon Aero pads used in other helmets including, especially the MICH, which he thought was "really bizarre." When he had tried Team Wendy pads in his own helmet, he found that they were very painful. After his helmet was in place for a while, points around his head really started to hurt. The pads were very hard, resembling packing material. Caldwell had read about the comfort-versus-protection issue, but he thought that if you can have both, why not?

At the end of the day, because of the Byzantine military procurement system, a system so complex and illogical that trying to navigate it would turn the strongest and most stable individual into an enraged maniac, the Oregon National Guard was unable to purchase Oregon Aero pads. Thus a decision had to be made whether to hang on to the earmark money and try to make a purchase of other pads or return the money to the National Guard Bureau.

Caldwell decided to return the earmark money because the Guard was not allowed to buy what the Army considered "unapproved pads." Since the money returned to the bureau was earmarked for pads, it sent the funds to the Kentucky CIF facility for the purchase of pads. The facility purchased Skydex pads, which, like Oregon Aero pads, were also unapproved by the Army. When he learned of the purchase, Caldwell was dismayed. He knew that the Kentucky facility was covered by the same procurement rules as the Oregon Guard. Caldwell wanted to know how it was possible for the Kentucky CIF to buy unapproved pads when the Oregon Guard could not.

He eventually learned that the Kentucky CIF facility had decided to go through the General Services Administration (GSA) to purchase pads with the returned earmark money. The military can, at its discretion, use the GSA to buy certain products. If a specific manufacturer is on the GSA schedule, it can bid on that product. A solicitation was issued electronically through its GSA E-Buy system, which allows government buyers to obtain quotes online from vendors.

The GSA solicitation, called a request for proposal (RFP), stated that the pad specification requirement must meet or exceed the Skydex product. The Federal Acquisition Regulations (FAR) allow a solicitation to include a brand-name product as the "brand name or equal" standard for submission of quotes. Five quotes were submitted in response to the Kentucky CIF facility's solicitation. Included in those quotes was one from a vendor that sold Oregon Aero pads. In its bid this prospective vendor did not address the testing data but stated that it had a better product that provided better shock attenuation and was more comfortable than the Skydex product. Its proposal did address a flammability issue with Skydex and acknowledged that it had a wearability problem.

Instead, GSA awarded the $1.8 million contract to the lowest bidder, Atlantic Diving Systems (ADS), located in Norfolk, Virginia, which sold all types of military equipment, and Skydex pads exclusively. According to sources ADS was politically connected with the military, especially with Special Forces. It had gained access to the market through military friendships it had established. The determination to procure the Skydex helmet pad kit from ADS was made based on price, knowledge of and previous experience with the products, an examination of the samples, and delivery.

GSA may have gone with the lowest bidder, but soldiers had complained of a wearability problem with Skydex pads. After they wore them for a time, the lining broke through, and a wiry mesh began to scrape the scalp.

Outraged at the Skydex purchase and that politics would win over soldier safety, Caldwell thought about making a federal case out of the Kentucky CIF purchase of unapproved pads, such as triggering a congressional or inspector general inquiry of the issue. However, if he did that, the Oregon National Guard could be punished in retaliation. Instead, he decided to complain directly to the NGB. Caldwell pointed out to the bureau that it had made a mistake, and asked, if it could buy unapproved pads, why couldn't he? The NGB quickly went into bureaucratic protection mode, saying only that it would look into the matter—an answer that is code for nonaction. The Kentucky CIF reports directly to the bureau, so Caldwell felt that complaining about its actions would likely be biting the hand that feeds you. The only response would most likely be that some low-level person who was not knowledgeable about the rules made a mistake, and it was now water under the bridge. The bureau was still not going to let Caldwell buy Oregon Aero pads. Evidently, according to Caldwell, in order to make sure soldiers were provided safe protective equipment, Army or Guard units had to go to private sources to get the equipment.

So they did. All of Caldwell's officers ended up buying Oregon Aero pads out of their own pockets and installing them in their helmets when they were deployed. Many of the enlisted men also bought the pads out of their pockets, with some going through Operation Helmet. In the end the military procurement system failed the troops of the Oregon National Guard in a time of war.

CHAPTER 23

Fallout

nitial media attention of the Sioux Manufacturing case was mostly limited to newspapers in North Dakota and based on an Associated Press report taken from both the Department of Justice and Sioux Manufacturing press releases. The local *Devils Lake Journal* added a quote from Sioux Manufacturing CEO Carl McKay, providing his explanation of the decision to settle the court action as a necessary business decision to limit legal costs and because "SMC likely would have lost some new orders by government entities and others because of the uncertainty caused by litigation. SMC is a small tribal-owned business that has limited financial resources."[1] The *Grand Forks Herald* quoted U.S. attorney Drew Wrigley: "They were underweaving. That is not debatable. It's apparently spinnable, but not debatable."[2]

Both Jeff Kenner and Tammy Elshaug didn't like what the local papers were saying, especially the *Devils Lake Journal*'s statement that the allegations were brought by "disgruntled employees."[3] Jeff felt the DOJ press release was "wishy-washy." He didn't like language such as "on occasion, SMC knowingly delivered cloth that had not been woven to precise specifications."[4] Jeff called Campanelli and expressed both his and Tammy's displeasure over the media accounts of the case.

Campanelli normally didn't publicize his *qui tam* cases in order to protect the relators. However, with both Jeff and Tammy upset over newspaper stories, he called *New York Times* reporter Bruce Lambert, someone he had dealings with in the past. Lambert, a twenty-two-year veteran reporter with the *Times*, was interested in doing a story, and Campanelli promised an exclusive. There was also talk of doing a series of three or four articles to run in the paper.

In addition to writing an article about the case in the *Times*, Lambert also contacted an acquaintance who was a producer for CBS's *48 Hours* to discuss doing a piece for the program. Campanelli's goal was to have at least ten million people see the story on television and read about it in the news. He wanted Jeff and Tammy to fly to New York, to be ready to appear on various television programs once the article appeared. Lambert believed the story would not end once it was printed in the *Times*. Campanelli envisioned contacting CBS, MSNBC, *Inside Edition*, the *Daily News*, and *Newsday*, along with many local television stations, once the *Times* story ran.

On February 6, 2008, the article appeared in the *New York Times* under the headline "Manufacturer in $2 Million Accord with U.S. on Deficient Kevlar in Military Helmets." Sioux Manufacturing CEO Carl McKay did not respond to Lambert's request for an interview, so Lambert quoted McKay's statements from the company's press release denying the allegations.

According to Gwynedd Thomas, an Auburn University professor specializing in ballistics and protective fabrics who was quoted in the article, "You must have a certain amount of protection, and you can't go below that." Thomas elaborated: "Although the difference between 34 and 35 threads per square inch seems modest, the cumulative loss in layers of fabric is significant. Every time that you're losing some mass, you're losing some integrity."[5]

In responding to the issue of Sioux Manufacturing having used extra resin to compensate for the reduced weight resulting from a shortage of Kevlar, Thomas noted, "Extra resin also poses a hazard to soldiers. If they were putting more resin in, they were doing something that will hurt soldiers, because it reduces elasticity and increased brittleness." Her opinion was that the 5 percent reduction in Kevlar fiber crossovers resulting from weaving thirty-four by thirty-four instead of thirty-five by thirty-five was "quite a lot." She expressed surprise that "somebody [was] not pursuing that more vigorously from the government," adding that if she were a soldier's parent, she "would want to give [her] son a better helmet."

As expected with an article in the *New York Times*, the fallout immediately began in Washington. On the same day the article appeared, the nonprofit advocacy organization Citizens for Responsibility and

Ethics in Washington (CREW) sent a letter to the chair and ranking member of the Senate Armed Services Committee demanding an investigation of the contract awarded to Sioux Manufacturing after the company was fined for providing faulty helmets to U.S. troops. In its letter CREW expressed "outrage" at learning that the DOD was "betraying its responsibility" to protecting the troops by awarding another contract to Sioux a "mere 12 days before the lawsuit was settled," with knowledge of the problem with the helmets and the "company's efforts to cover it up."[6]

Within a few days of the *New York Times* article, then senators Hillary Clinton and John Kerry issued press releases calling for an investigation of the Kevlar case. Senator John Kerry sent a letter to Department of Defense Inspector General Charles Kicklighter requesting a report on "sub-standard armor inside protective helmets that were distributed to troops over the last twenty years, including those fighting in Iraq and Afghanistan." He added, "The possibility that a company knowingly put troops at greater risk is unacceptable, but reports of the Pentagon's inadequate response once these allegations surfaced are even more troubling." Kerry further requested that Kicklighter "conduct a thorough review of this matter."[7]

Considering that the *Times* was widely read in policy-making circles in Washington DC, including the Pentagon, Sioux Manufacturing and the tribal council felt that a response to the article was necessary. On February 12 the chairwoman of the Spirit Lake Sioux Tribe, Myra Pearson, sent a letter to CREW "strongly protesting" the *Times* article along with a "sentence-by-sentence rebuttal" titled "Commentary by Spirit Lake Tribe and Sioux Manufacturing Corporation concerning *New York Times* Article of February 6, 2008." The commentary was an attempt to repair their reputations, but it clearly revealed their contempt for the investigation, the DOJ, and the relators and their anxiety over the bad press. At the same time, the tribe's attorneys issued a press release to the media regarding their protest. Pearson complained that the *Times* article was not only "riddled with inaccuracies, but also highly display[ed] a stunning lack of factual research and journalistic professionalism."[8]

Pearson also expressed concern that several members of Congress had called for an investigation, and "many soldiers and their family members ha[d] been needlessly frightened by false assertions that the helmets [were] unsafe, and SMC's reputation ha[d] been wrongfully injured."[9] Letters were also sent to members of Congress, including the North Dakota congressional delegation, and groups like CREW that had expressed concern.

Much like its answer to the complaint and press releases, Sioux Manufacturing, in its commentary, strongly denied weaving Kevlar cloth below the thirty-five-by-thirty-five specification. Additionally, it maintained that the monetary settlement was diminished to $2 million "predicated on the government's recognition that many of the asserted 'facts' and 'claims' presented by the 'whistleblowers' were not founded on fact or law."[10]

In response to U.S. attorney Drew Wrigley's statement that it was not debatable that the company was underweaving, the commentary claimed that Wrigley's opinion *"was not credible."* His statement, the commentary asserted, was "further impeached by the fact that during the entire course of the investigation the government did not locate one piece of non-compliant cloth." Additionally, Sioux Manufacturing stated that the "most disturbing element of this matter" was that the DOJ and the DOD accepted the allegations of the relators, who were, "fired for inappropriate conduct on company time in a company vehicle, as gospel, without any skepticism of their motivations or a careful review of the solidity of their legal theories or facts."

Then Sioux Manufacturing took their shot at Campanelli, saying he "is an attorney who advertises on the web that he can get plaintiffs 'huge monetary awards.' His web site lists dozens of fraud cases against government contractors, which resulted in money being paid to whistleblowers, and he implies those cases were brought by his firm, which is not true." Also, Carl McKay denied that a conversation occurred between him and Jeff Kenner regarding the underweaving problem. He contended that "neither Mr. Kenner nor Ms. Elshaug 'objected' to SMC, the Tribe, Mr. McKay or Mr. Dauphinais about any allegations or suspicions that they had of 'fraud' prior to their being terminated."

Despite the DOJ investigation, which found no evidence of an affair between Jeff and Tammy, Sioux Manufacturing, in its effort to discredit both of them, continued to claim that they were terminated for violating company policy: "Although both were married, they appeared to have an inappropriate relationship with each other, rather than a business relationship. This conduct was tolerated, although they were advised that this conduct was not proper." The commentary continued with a narrative of their sighting at a remote location by fire department personnel that embellished the written statement provided by the fire chief who was at the scene.

Sioux Manufacturing again insisted in its commentary that it followed the "minimum standard" for weaving Kevlar cloth. However, it pointed out that the military had "proposed a revised minimum standard for Kevlar weave of 34 x 34 yarns per square inch."[11] It is true that in 2008 the Army did propose reducing the weave-density requirement without considering the effects of a weakened Kevlar at thirty-four by thirty-four yarns per square inch. According to an Army Research Laboratory (ARL) official, the change was actually proposed at the suggestion of industry. "Industry suggested it and ARL put it in the revision," he explained but claimed he didn't know why industries wanted it downgraded.[12] An official at the Philadelphia Defense Supply Center, which prepared the official revision, agreed with the ARL that industry recommended the downgrade from the original specification in 1981.[13]

The ARL official explained the process for the proposed weave-density change: "The way it works is that industry supplies ARL with what they say they can make it with. If you've got multiple industries, they go to different companies. They are not going to just take the word of one. ARL does not have the funding to do their own testing. They have to rely more on industry. But ARL still makes sure what industry gives them is what's best for the soldier."[14]

Not doing the testing of the proposed downgrade of the weave density and relying only on industry did not sit well with the Navy. When the proposed downgrade was submitted to the military departments for review and approval, the Navy reviewing activity immediately questioned the change and the rationale. As a result the recommended change was rejected, overruled, and removed from the draft prior to finalizing the published edition in February 2009. The weave density remained at thirty-five by thirty-five yarns per inch.

Campanelli thought the Sioux Manufacturing response to the *Times* article was "pretty disgraceful, a spin document that clearly was prepared by an attorney. It wasn't a lay person writing it." He said, "Short of pointing one to a fairy tale, it would be difficult to identify a less credible work of fiction. . . . Some of the stuff in it goes from ludicrous to ridiculous."[15] However, neither the Sioux Manufacturing commentary nor Campanelli's response were printed in the *New York Times*, but they did appear in some local North Dakota newspapers.

By the end of February, members of the congressional delegation from North Dakota had weighed in on the case. Like his colleagues Clinton and Kerry in the Senate, Senator Byron Dorgan (D-ND) on February 28 called for a Pentagon investigation "into Kevlar safety and allegations against the Sioux Manufacturing company." However, unlike his colleagues, he indirectly questioned the investigation, saying, "If the investigation determines that the facts do not support the allegations, then it raises other questions about how and why this matter was pursued."[16] Dorgan was chairman of the Democratic Policy Committee, which was one of the few voices in the Senate opposing contractor abuses in Iraq, and he had held a series of hearings on the issue during 2003–6, when the Republicans were in control of the Senate.

David Peterson was not happy about Dorgan's comments calling into question whether his investigation was accurate, but there was nothing he could do about it. DOJ regulations did not permit him to comment on Senator Dorgan's statements or contact him. He had known Dorgan for over thirty years and felt that he tried to do the right thing, but he could not talk to him while Peterson was still with the U.S. attorney's office. He believed that Dorgan got his information on the case from his Democratic colleagues in North Dakota, who were representing the tribe. Peterson strongly felt what Dorgan was hearing and putting out to the press was not accurate based on information in the case.[17]

On the same day that Senator Dorgan made his statement about the case, CREW announced it had joined VoteVets.org, a public action committee whose mission is to elect Iraq and Afghanistan war veterans to public office, to call for an investigation by the Senate and

House Armed Services Committees of the DOD contract awarded to Sioux Manufacturing. CREW also announced that it had posted on its website the tape recordings that Jeff Kenner made of Sioux Manufacturing employees "discussing the production of substandard Kevlar for helmets." The executive director of CREW, Melanie Sloan, asked: "How was Sioux Manufacturing allowed to make insufficient Kevlar in the first place and then once that was exposed, how could the company be awarded a new contract? Are there soldiers who would have escaped injury or death if their helmets had met government standards rather than the inferior ones produced by Sioux? Frankly, the chairs and ranking members of Armed Services, Senators Carl Levin and John McCain and Representatives Ike Skelton and Duncan Hunter cannot be permitted to take a pass on this."[18]

CHAPTER 24

Delamination

I n July 2007, more than a year after Assistant U.S. Attorney David Peterson requested that the DOD conduct ballistic tests of PASGT helmets, the tests were finally conducted by the ARL at Aberdeen Proving Grounds, Maryland. The testing followed the standard military specification for helmet ballistic tests to determine whether a projectile could completely penetrate the helmet.

The PASGT helmets used in the tests were obtained from the DOD supply system and purportedly manufactured by UNICOR. The goal of the tests was to determine the velocity at which 50 percent of the projectiles fired against the helmet were expected to completely perforate that helmet. Late in July 2007, the ARL reported that the helmets tested met the ballistic protection requirement.[1]

However, focusing the testing only on complete projectile penetration did not completely answer the question of whether a Kevlar weakened from shorting the weave could have caused brain injury. When a projectile of sufficient velocity impacts the helmet but does not completely penetrate, kinetic energy from the projectile can cause the interior of the helmet to delaminate or deform. This phenomenon creates an interior bulge that can exceed the standoff distance of the half-inch between the helmet shell with a sling (padless) suspension system and the skull. The bulge is called backface deformation (BFD). It has been known as one of the primary damage mechanisms of Kevlar helmets. BFD was routinely tested on Kevlar body armor but not on Kevlar helmets. The impact to the head caused by BFD is a possible cause of skull fracture and brain damage leading to TBI. There was no evidence that ARL considered BFD as part of its ballistic testing of the PASGT helmets. If it had tested for BFD, the results may have changed the dynamics of the Sioux Manufacturing case.

Canadian scientists Jack van Hoof and Michael Worswick, at the University of Waterloo in Ontario, studied the effect of BFD on the head of a soldier in 1999–2000. Their research was conducted with the intent of developing better Kevlar combat helmets for the Canadian armed forces that were using the PASGT helmet at the time. They determined that even if a Kevlar helmet was able to stop a projectile from complete penetration, death or a severe head injury could occur if delamination or deformation on the interior of the helmet was too great. Worswick reported that "a Kevlar helmet works by 'catching' a projectile and absorbing its energy. However, the faster the projectile is moving when it hits the helmet, the more likely it is to cause deformation and delamination of the layers of woven Kevlar. If this deformation bends the inner part of the helmet too far, it will of course collide with the skull of the wearer."[2]

The scientists noted that the standards for testing ballistic helmets ignored the deformation of the laminate, focusing instead only on the prevention of complete projectile penetration. Van Hoof and Worswick felt that the "deformation of the composite is of vital importance." Among their findings they noted that fiber breakage in both weave directions was one of the "principal damage mechanisms in ballistically impacted laminates."[3]

The National Academies of Sciences, Engineering, and Medicine reported a study of BFD in 2003 demonstrating that BFD resulted in skull fractures. Researchers "performed a study to develop injury criteria for skull fracture in human heads during ballistic loading of a protective helmet.... These ballistic impact tests were used to assess the risk of skull fracture and other head injuries from nonpenetrating BABT [behind-armor blunt trauma] for military helmets. Skull fractures were produced in five of nine tests, with a single artifactual fracture at a preexisting unhealed craniotomy. Injuries ranged from simple linear fractures to complex combinations of linear fractures and a depressed fracture."[4]

DOD testing expert Pierre Sprey, after reviewing the Sioux Manufacturing case, also felt that the DOD failed to account for the possibility of delamination or enhanced delamination of the PASGT because the Kevlar was weakened from shorting the amount of fibers in the weave. He believed that the ARL testing was useless. The fact that the ballistic test specifications at that time only tested for projectile penetration versus no penetration only demonstrated the incompetence of the Army's helmet specification writers.[5]

Sprey gave an example of the incompetence that was rampant in military testing. He recalled a testing incident of squad-level radios. The Army conducted fraudulent jungle tests in Panama to prove that the radios worked in the jungle. They went to the very edge of the jungle, cleared a path to where the receiving antenna was located—a clear line of sight—and the radio worked. This unrealistic testing did not show the flaws in the radios and allowed them to be fielded in a real combat situation, where they failed. This became evident during the early 1990s when squad-level radios often did not work when most needed. It was one of the reasons for the disaster in Somalia with Blackhawk Down, the crashing of several U.S. helicopters and the death of a number of soldiers. The soldiers had Motorola radios that failed to communicate from ground to helicopter or to their base. Thus they could not coordinate their attacks and were sitting ducks.

Sprey was adamant that "there is simply no doubt that, whether or not the helmet is penetrated, less Kevlar fibers per inch means greater deformation or delamination on the inside of the helmet shell. That, in turn, means greater damage to the brain. The only open question is how much that increased damage amounts to, given a 5 percent reduction in weave density [alleged in the Sioux manufacturing case]." Importantly, he felt that "a 5 percent weave density deficiency would create a huge increase in brain trauma [from a bullet/fragment impact] because the 5 percent deficiency [would] certainly increase the maximum deflection/delamination enough to allow a highly damaging impact with the skull."

Sprey believed the real tragedy was that helmet tests designed to record maximum delamination were "so easy and cheap to do and so obviously critical to saving the lives of soldiers." The Army did not conduct them because of the "incompetence of the helmet bureaucracy" and the possibility that the test results might interfere with "this venal bureaucracy's ability to award contracts to their favorite manufacturers."

In explaining the effect of shorting the weave of Kevlar cloth, Sprey pointed out that fewer threads per inch means less tensile strength in each layer, causing a stretching or elongation of the fibers pulling on the outer parts of the layers. This causes more relative motion between the layers, leading to a greater tendency to delaminate. Woven Kevlar fabric is stable only as long as the fibers are in tension. The outer parts of the layers, according to research, carry the greatest loads and therefore will fail first with fiber failure moving inward.

Eventually, the Army's helmet bureaucracy became concerned about the effects of helmet BFD on a soldier's head. Awareness of and testing methodology for body-armor deformation already existed, but the helmet bureaucracy was slow to apply to the helmet what it had learned from body-armor testing. The bureaucracy finally became aware of the problem years after TBI became the "signature injury" of the wars in Afghanistan and Iraq.

In January 2009, more than six years after the start of combat in Afghanistan and more than five years after the initial invasion of Iraq, research began at the ARL into an efficient and reliable way to measure helmet BFD and explore its possible correlation to injury. Dixie Hisley, an engineer for the Experimentation Branch of the Survivability Lethality Analysis Directorate at the ARL, led this effort. She admitted in an email that it was "well known that helmet BFD may potentially impact a Soldier's head, thus [her] focus [was] to understand how to measure that impact in an experimental setting and provide the data to others who may provide correlation to injury." She agreed with Canadian scientist Jack van Hoof that BFD "is considered to be a possible cause of skull fracture and underlying brain damage."[6]

The result of her effort "has shown that helmet BFD does load the skull similar to blunt-object impact and DIC (Digital Image Correlation) instrumentation and methodology permits robust, repeatable, dynamic helmet behind-armor measurement data to be collected." However, she deferred to the helmet bureaucracy at PEO Soldier to decide whether the techniques shown in her research are used in helmet standards or acceptance tests.

Despite ballistic-impact studies of the PASGT helmet conducted internationally, the Army helmet bureaucracy failed to acknowledge or consider those studies for years. It apparently failed to conduct its own research and testing to determine the effects of Kevlar fiber weakening on the extent of delamination and possible head injuries as a result. It also failed to consider delamination in the testing of the PASGT helmet conducted in July 2007 in response to the DOJ investigation of Sioux Manufacturing. This failure left open the question of the true effects of weakening the Kevlar fiber structure when the weaving process was shorted.

CHAPTER 25

Won't Work Unless You Wear It

By 2008 Operation Helmet had sent more than thirty-eight thousand pad kits to the troops in Iraq and Afghanistan. Doc Bob learned from many of the troops requesting the pads that wearability of the Army's ACH and the Marine Corps's LWH with stiffer supply-issued pads was becoming a serious problem. Many marines were exchanging the supply-issued pads for Oregon Aero pads and ignoring the Marine Corps's policy of using only the supply-issued pads. Justin Meaders told his grandfather that the LWH with the supply-issued pads was trash. He couldn't squeeze it down on his head. It was like having a Brillo pad inside his helmet.

Many of the emails to Doc Bob cited discomfort resulting from the stiffer supply-issued pads, including headaches that were affecting mission and readiness. "Essentially, we go on resupply convoys through rough, rocky terrain, sometimes driving for up to 20 hours straight on a drive that can last up to 3 days," a marine wrote to Doc Bob.

> The level of concentration and alertness needed to make it through some of this terrain and not get a vehicle stuck is pretty high. . . . For some Marines in our platoon, their helmet pads have deteriorated to the point where their head is almost against the helmet itself. . . . The pads issued to replace the leather bands are so stiff, they couldn't even get the helmet on their heads and put the leather band back in. This discomfort causes headaches, and it requires more energy to concentrate on the road or terrain instead of our heads.[1]

Incredibly, the Army and the Marine Corps continued to stress to Doc Bob that they never received any complaints about the helmets. In frustration Doc Bob, through his website, urged troops to report the problems with the new pads to Army and Marine Corps leadership. In reality the military culture does not encourage soldier or marine complaints up the chain of command. "If there was a problem with the helmet and helmet pads causing headaches or other problems, they just do not complain to higher authority," Justin Meaders revealed. "Marine Corps officials are not getting the complaints," he said. "It was more the small unit leadership who were trained to stop complaints at that level. They only take necessary information to push up the chain. If someone comes to unit leadership with a problem they can handle, why should it go any further?"[2]

Justin described the situation from his perspective: "Guys are going to their unit leaders, like corporals or sergeants, maybe even staff NCOS (noncommissioned officers), and saying it's messed up and the reply would be 'you're a fuck'n Marine, suck it up.' You can voice your opinion, but it doesn't mean it will go anywhere. Jumping the chain of command in the Marine Corps is a big, big deal. You do not go above the next guy. If you do, you had better have some weight behind it or your fitness report is going to be bad."

Another problem that inhibited complaints from Army soldiers is that most have never heard of PEO Soldier, let alone complain to it. "This is the first time I have heard of PEO," an Army operations officer wrote Doc Bob. "The reason they have not had an official complaint is probably because not many people know the website exists. . . . I'm trying to find information on any item and nothing is showing up. . . . It's not much help. . . . If they're not interested in feedback, then they're on the right track and need to do nothing." After Doc Bob's notice to his contacts within the Senate Armed Services Committee of this dilemma, a committee staffer tried to enter a complaint about the helmet pads with PEO Soldier in 2010 but was unable to do so.[3]

Also, Doc Bob's attempt to report the complaints to PEO Soldier was met with a dismissive attitude by the project manager for testing. He told Doc Bob that those who switched from the old PASGT helmet to the new ACH with Team Wendy pads thought the new helmet was wonderful. Those who had only switched from Oregon Aero pads to Team Wendy pads were those who bitched. He felt that anything you gave a soldier to put on his head would cause complaints. He held to the line that helmets were for protection, not comfort. However, according to Doc Bob, he ignored the fact that the major issue was not just discomfort or head pain but distraction from combat duties.[4]

Wearability factors such as comfort are important in determining whether soldiers or marines will wear their helmets properly or not at all. When wearability problems reduce the use or the proper use of helmets in a combat zone, they increase the troops' risk of head injury and TBI.

Oregon Aero CEO Mike Dennis explained, "When you try to interface something to a person, it's a much narrower band of choices and a much tougher job. Human beings are the most difficult packaging job on the planet. You've got to breathe, you got heat issues, contact issues, circulatory issues, and you got to function. You have to account for all different types, shapes, genders, and doing different tasks."[5]

Mike has seen representatives of U.S. government laboratories stand before a camera and say soldiers like to hurt; it keeps them sharp. He feels that soldiers are no different than anyone else, that no one likes hurting, and it doesn't keep people sharp. They focus on the cause of the pain to the exclusion of the task they started out to do. "If you are sitting in pain," Mike continued, "you are not a good driver, you are not a good pilot, you are not a good secretary, it's a distraction. The Air Force caught on to the fact that a good seat cushion allows the pilot to put more bombs and bullets on target. The numbers went up and the Air Force was real happy about that."

Like the Army helmet bureaucracy, the Marine Corps's helmet bureaucracy also maintained that there were no complaints from marines in combat about their helmets. Despite their denials Doc Bob continued to receive email from marines expressing their displeasure with the supply-issued pads. One such complaint from a deployed Marine Corps lieutenant explained to Doc Bob the reality of helmets with supply-issued pads being worn in combat:

Wearing a Kevlar [helmet] for more than a couple hours is probably one of the more terrible experiences Marines have to undergo. I'm sure that my Marines would all agree that more comfortable pads would definitely lead to better performance during missions, less fatigue, and fewer headaches. This request is characterized by a desire to have a better product to allow my Marines to perform to their maximum capability. I understand that even the most comfortable pads will lead to fatigue, headaches, and pain. I don't have any facts or figures to back it up, but I think that if the pads are safe and comfortable, it beats simply being safe.[6]

By 2008 TBI cases were at an all-time high for troops fighting in Iraq and Afghanistan. During that time Doc Bob was inundated with complaints from marines and soldiers deployed to Iraq. In turn he continued his campaign to get the Army and the Marine Corps to focus on the wearability factors of pad systems and its connection to TBI risk. Doc Bob wrote congressional and military leaders, citing complaints from combat troops, urging them to rethink the purchase of cheaper, poorly engineered pads.

In order to obtain a much better indication of wearability issues for deployed marines and soldiers, Doc Bob and his son, Mark, initiated a troop feedback survey in late 2006 administered through the Operation Helmet website. The survey was designed to determine the comfort and satisfaction levels of combat helmet-pad suspension systems. It was an improvised, unscientific survey, but it at least provided a good indication of how the troops felt about their helmet pads. The data, accumulated over a four-month period, showed a clear preference for using Oregon Aero pads by those who had worn multiple pad suspension systems because the Oregon Aero pads provided good levels of subjective comfort. Of all the responses naming a specific pad system, Oregon Aero's was rated the highest followed by Skydex and Team Wendy.[7]

Interestingly, the 2005 performance specifications for the ACH clearly stated that "materials used in the retention system shall be suitable for use, including prolonged skin contact, and be comfortable." Also, "the pads shall . . . provide standoff, comfort, protection, and stability."[8] The specification mentioned comfort fourteen times. In 2011 a revised specification mentioned comfort only three times. "They realized comfort would give them a black eye so they just quit testing for it," Doc Bob commented.[9]

The original requirements for the pads called for submerging them under sixty-six feet of saltwater for twelve hours. After removing the pads and weighing them, the analysts compared this weight to the preemergent weight to see if the pads had gained weight underwater. The newer version of the test called for submerging the pads in only six inches of water for twelve hours. The pads were dried with paper towels then air-dried for twenty-four hours to determine if they were waterproof. "It's like testing deodorants," Doc Bob said sarcastically, "by having some sweaty guy, after a workout, try it on and then take a bath and see if he still stinks afterwards."

After numerous complaints to Doc Bob from marines in combat and his attempts, over a ten-year period, to convince the Marine Corps equipment bureaucracy that the helmet pads in the LWH had wearability problems, the Marine Corps held an "Improved Helmet Suspension System Industry Day" in October 2012. It stressed the primary importance of headgear wearability, noting that "wearability influences long-term effectiveness of warfighter."[10] However, it continued to supply the same pads that were the source of the complaints.

Research conducted at John Hopkins and published in 2011 by the *Journal of the International Headache Society* concluded that headaches constituted a significant burden to military units deployed to combat zones. Study leader Steven P. Cohen explained that most headaches were the result of damage or pressure on the back of the head caused by Kevlar helmets worn by troops on patrol for long periods of time. He reported, "Everyone who goes on patrol wears a Kevlar helmet. . . . If you get a headache from your helmet, you still must wear it. If you can't tolerate your helmet, you can't do your job. It would be too dangerous. So these folks end up being evacuated and not returning to duty."[11]

The study found that headaches became one of the "fastest-growing causes of medical evacuations of the two prolonged military conflicts." Yet evacuation was just the beginning of the problem. About two-thirds of those evacuated for headache problems never returned to duty. The Marine Corps, the study found, had the highest rate of combat troops not returning to duty.

The complaints about headaches that Doc Bob received from troops in combat, consistent over the years, reflected the John Hopkins study findings, as the following excerpts from eight emails to him illustrate:

My current pads are rock hard and over any extended period of time they cause headaches and pain. My old pads make it nearly impossible to wear any helmet for longer than 10 or 15 minutes. After that I would remove my head protection to work more smoothly. I badly need new pads to help me think straight and allow me to wear my helmet for longer periods of time.

The current helmet pads have many issues that affect mission performance. They cause terrible migraine headaches. The last thing you need to be doing after returning from a patrol that often times is several hours long is dealing with a migraine. Also this is an issue while standing post, being distracted by a migraine while you need to be as alert as possible is another unnecessary problem.

The current pads that we have issued to us give just about every Marine headaches and that is the last thing that we need when going on patrols over in Afghanistan.

The headaches affect us in that we are constantly re-adjusting our helmets while on patrol trying to make it as comfortable as possible. This in itself seems small however it takes our focus off the matter at hand on patrol sometimes, which is looking out for threats. The headaches sometimes become so intense that they cause unwanted squinting and in some cases have been noted to temporarily affect vision.

I have more than occasionally found my soldiers with kevlars [helmets] off while on mission, specifically due to pain on the head or eventually headaches.

Every day I wear my helmet for at least 8–10 hours when outside the wire. I have to wear them on dismounts to keep myself safe, even though they give me terrible headaches because of the constant pressure each "pad" puts on my head. I'll admit I do not always wear my helmet in our armored vehicles, because it tends to hurt so bad, even though when I'm inside the vehicle is one of the most important times to wear it due to IED's.

It is very difficult to concentrate when we're on patrol and sweeping when the current pads we have give us all headaches. They take away from situational awareness.

On some of the missions they are in their trucks for up to 14 hours and the pads we currently have caused headaches, and pain from the pressure on their heads. The undue discomfort causes the Soldiers to remove their helmet to get some relief. My Soldiers deserve to have a helmet that they can wear without causing pain and also provide protection from blast injuries.[12]

During the summer of 2008, the DOD's primary testing agency, the DOT&E, conducted its own user evaluation and assessment of combat helmet-pad suspension systems, mandated by the fiscal year 2008 National Defense Authorization Act. Instead of testing under actual combat conditions, it used new recruits at military training sites within the United States, at Fort Benning, Georgia, for the Army, and at Quantico, Virginia, for the Marine Corps. Four padded suspension systems were tested: Team Wendy, MSA, Oregon Aero, and Skydex.

Not surprisingly, given the more sterile conditions of training facilities, the results from user data collected through use and questionnaires reported "no significant difficulty in completing military tasks with the pad systems," and interestingly, "no significant wearability issues" and "no significant differences . . . between the pad systems." Therefore, DOT&E concluded, "There is no compelling reason, based upon these evaluations, to change current fielding plans for either Service."[13]

Defense analyst Pierre Sprey weighed in on the DOT&E user evaluation. He was blunt in his opinion. After reviewing the user report, he commented,

It is hogwash. Total hogwash. It was a disgrace of a test. It was a questionnaire survey. Can you imagine an issue that is as important as whether a guy is going to keep his helmet on after ten hours? They go down to some peacetime base and they issue the helmet to a few hundred soldiers and let them carry it around for a week then give them a questionnaire afterwards? What's worse, in their summary report, they don't even say what the questions were on the survey. If they were political pollsters, they would be out of business within 24 hours for denying you access to the questions. Questions that tell you how they judge the percentage of people who think that one isn't different than the other.[14]

Sprey offered suggestions for designing a user evaluation test:

The right way to do that test is so simple. You devise two scenarios. One is a ten-hour march with combat gear and every couple of hours drop into firing position and do some firing exercise, then march on. That's one scenario. You put a helmet on a guy and let him go through this scenario. The next day, give him a helmet with a different pad and run him through the same scenario. Then you take detailed notes of what happened. Where you able to fire as well with the new pad? Did you get a headache? How was the sweat situation? Does it smell bad at the end of the day? That's the way you do a serious comfort test. Comfort is a question of combat effectiveness. This is not a comfort-like luxury. This is comfort like living or dying.

Sprey continued, "Then you do another scenario of the same thing inside of a Humvee or inside a Bradley Armored Vehicle and put them out there for 10 hours the same way. He has to sit in the vehicle, ride in the vehicle, stop, run out, go through a firing exercise, jump back into the vehicle, go through it three or four times during the day, and then at the end of the day you see what happened. Did the helmet slip a little? Where you able to do the firing exercises and aim fast enough? Did you get a headache? Could you hear?"

"That's the way you do a serious comfort test if you really care about the soldier and care about keeping him alive," Sprey advised. "If you are just trying to paper over something, then you issue it to a bunch of bored, peacetime troops who are standing around some base somewhere and going through their standard training and for whom the whole helmet issue is a big pain in the butt. Then, at the end of the week, they get a big questionnaire they don't understand with some confusing verbiage on it, of course they aren't going to see differences."

"They don't have to use soldiers in combat, but those in combat training and put them in a combat scene. That's what good operational testing is all about. That's what I have spent half my career on. It's not hard. It's not rocket science. You just have to care, think about combat, and not defending the prerogatives of your bureaucracy. Given what I saw of the user evaluations, I wouldn't even have taken the effort to throw them in the wastebasket, they are so useless."

Doc Bob also denounced the DOD's Limited User Evaluation, "where they put the helmets on for two whole days and their head doesn't hurt so they must be okay." "They do them at training facilities," he continued.

> They bring out a guy who had just been in training and had been wearing a PASGT during the training and then he gets a new padded LWH. He doesn't care what the pad is; it's paradise compared to the PASGT. He wears each pad for two days, but that's not long enough to tell anything. You don't strap it on from nine to four, daytime patrol or simulated combat. How the Perfumed Princes of the Pentagon [and Belvoir and Quantico] can claim this short-term user evaluation is equal to reports from troops wearing helmets all day, every day, for months on end is beyond reckless.[15]

A study on soldier satisfaction with both the ACH and the PASGT was reported by the publication *Military Medicine* in 2007. The study supported both Doc Bob and Pierre Sprey, calling for more rigorous and realistic survey methods of wearability factors. The study pointed out that "the methods used to evaluate helmets play an important role in the quality and usefulness of the information that is produced," and that such research requires testing "about soldiers' real-life experiences with helmets," in order to properly assess a helmet's overall protection and wearability. An important point in the article was that helmet developers "must make tradeoffs" between protection and wearability factors. According to the article, increasing the amount of ballistic protection at the expense of wearability would result in the helmet not being worn as often as it should. "When problematic human use factors reduce helmet use, military personnel increase their risk of sustaining brain injuries because a helmet cannot protect against injury when it is not used."[16]

But according to Team Wendy, the primary provider of the supply-issued pads used in the ACH and the LWH since 2005, protection was its number one priority in the engineering of the pads. However, its view of the comfort issue appeared on its website: "If it's not comfortable, it's your own damn fault." Doc Bob countered in his blog, "No it isn't. It's the fault of warriors in air-conditioned foxholes and slick leather chairs who believe with all their dark little hearts they know better than combat troops what works on the ground in combat."[17]

Another response to Team Wendy's ad appeared in the blog of Stand for the Troops (SFTT), a nonprofit educational foundation:

> Frankly, we are not sure what the message is, but we are quite sure that Wendy's management must be a little red-faced at the suggestion that men and women in combat are responsible for the ill-fitting and uncomfortable military helmets that they manufacture. After all, the Wendy Epic helmet is standard government issue and our military leadership surely knows what is best for the troops that serve in harm's way. It is interesting to note, that Team Wendy goes on to promote the safety features of their combat helmet. I suppose that might prove reassuring if troops were actually wearing the helmet, but one suspects that an ill fitting helmet may well be used for other purposes than protecting our troops from head injuries or even worse. Surely, Team Wendy can't be blaming the troops in the field for the uncomfortable helmet it produces? SFTT hopes that this is not the message you are trying to convey to troops that wear your protective gear.[18]

A 2007 *Army Times* article reported that marines complained their helmets with Team Wendy pads fit poorly and were so stiff that they were causing headaches. The article also highlighted Operation Helmet's efforts to provide a more comfortable pad system.[19]

John L. Sweeny was the Team Wendy CEO during this controversy. Angered by the *Army Times* article, he responded in a letter to the editor criticizing Operation Helmet for making misleading assertions in the article. He claimed his pads were not inherently uncomfortable and that "the article focuses on a few complaints of discomfort reported by Operation Helmet, the non-profit organization that promotes the use of the competitor Oregon Aero pad suspension system, despite extensive scientific testing by the Army's Aeromedical Research Laboratory which concluded that Oregon Aero did not make the most protective pad for U.S. soldiers."[20]

Claiming that hundreds of thousands of Team Wendy pad systems were used by deployed Army and Marine Corps units and "the services report no issues," Sweeny concluded "the argument made by Operation Helmet that the Team Wendy pad is inherently uncomfortable is false." He placed the blame for an uncomfortable helmet more, "on the size of the helmet shell than to the pad system."

In 2011 Doc Bob asked a soldier in Afghanistan who had complained about the discomfort of his helmet, if the helmet was properly fitted, would it eliminate the discomfort? The soldier answered that although helmets were made in different sizes, all the pads were the same, so regardless if the helmet fit right, the discomfort came from the pads, not the helmet size. In addition another soldier reported to Doc Bob that some of the soldiers in his unit had to remove almost all the pads from their helmets to relieve them of headaches, and this was "while they ha[d] the proper helmet size and all."[21]

Team Wendy, a Cleveland, Ohio–based pad manufacturer, was founded in honor of the founder's daughter, Wendy Moore, who tragically died in 1997 from a traumatic brain injury suffered while skiing without a helmet. Her family established the company to raise helmet awareness for recreational sports enthusiasts. Their original helmet padding was used in helmets for skiers, snowboarders, and participants in other recreational activities. Much like Oregon Aero's Mike Dennis, Dan Moore, Wendy's father, was considered a "serial entrepreneur, tinkerer, and engineer," and he personally developed a polystyrene helmet liner used in a ski helmet. It wasn't until 2004 that Team Wendy transitioned to making pads for military combat helmets.

"This is not coddling the troops," Pierre Sprey said, commenting on helmet wearability.

This is absolutely of the essence. If they force you out there to march for 10 hours up an Afghanistan mountain, you wouldn't keep that helmet on with those [stiff] pads in them. Danger always seems a little remote. It is very hard to stay on the edge, expecting an ambush in the next 15 seconds for ten hours. Nobody can do that. You would go crazy if you did. If you had a comfortable helmet, it is not an issue. People are going to leave it on because it is a good idea. If it's giving you a headache, you are going to say I can't stand it, I'm going to take my chances for the next few minutes and get it off my head.[22]

On that point a 2010 USAARL photo survey in Iraq and Afghanistan revealed that roughly half the soldiers in the field were not wearing their ACH or PASGT helmets properly, thus exposing themselves to an increased risk of TBI.

The comfort wars among pad manufacturers reached the television airways in 2009. In May 2009 *Dan Rather Reports* on the HD Network devoted the show to the controversy surrounding the comfort versus protection issue between Oregon Aero and Team Wendy pads. The founders of both companies along with Doc Bob were interviewed.

Oregon Aero's Mike Dennis explained that his pads contained a composite foam product coated in a very proprietary process that his company invented. "We build it here in this building," Mike told Rather, "and it breathes air. It resists water. Actually stops water. And yet it's cool, cool to the touch, cool to the wear. So we accomplished all our goals.... It's totally malleable. You touch it and it senses your body temperature and heat and pressure and it just absorbs all that and fits you perfectly."[23]

Dan Moore, Team Wendy founder, responded by blasting Oregon Aero, saying its helmets did not perform: "I mean they don't have the impact absorbing characteristics of our pads. They are unsafe to send into the field," he claimed. Moore felt that Oregon Aero pads were too soft: "It first was a complicated way to make it," Moore explained, "and it was the wrong kind of energy-absorbing material. It had to be stiffer. It had to have more energy-absorbing characteristics. And it was obvious to me that that was a problem."

Dan Rather commented that Moore said his theory was "proven out by the fact that his pads [did] best on the military's blunt-impact tests." Moore claimed that he had never received any complaints from the military, that the Army did not think there was an issue with comfort.

Rather then referred to Doc Bob and remarked, "He believes the Pentagon was humiliated by his efforts and decided, through a long bureaucratic procurement process . . . to do it their way. Despite the Pentagon getting the final say, he hears all the time from troops who aren't happy with their pads."

Doc Bob responded,

Basically all say when I go out on patrol halfway through the patrol, I've got to stop, take a Tylenol. I've got such a bad headache. And I'm an up gunner on a Humvee and I'm looking for snipers. I'm also watching the road. And intermittently, I have to duck back down inside the vehicle and take my helmet off. In the meantime that part of the machinery is blind. And it's dangerous. I don't want to do it. . . . And when I sent them Oregon Aero pads, I got that e-mail from him. It said, Doc, these pads are wonderful. It says I forget I have a helmet on. I do my job.

"First," Doc Bob continued, "let's start with the Oregon Aero pads that they are a foam sandwich with a comfort foam that will accommodate the little irregularities in your scalp. Kinda like these mattresses you read about for your body molds to it. And then it has a stiffer layer of foam that slows down the impact between your head and the helmet." On the other hand, he added, "The Team Wendy pad is encased in plastic and has two different layers of foam in it as well. But if you feel those two different layers of foam, you see there is a marked difference."

Doc Bob's grandson, Justin Meaders, weighed in, telling Rather,

Those helmets that we had, we had the ones with the leather strap when we were about to deploy and it was trash. It gets all sweaty. It starts sliding around on your head when you put the NVGS on, mount 'em to the helmet, it'd fall down and you're sittin' there with one hand on your weapon and one hand on your helmet trying to hold them up. And it's just . . . just wanted everyone to be safe and have the same advantage. If it's not comfortable, I'm not gonna wear it. I mean I'll wear it because they tell me I have to wear it, but I'm not gonna be happy about it.

Afterward, referring to the Rather piece, Mike Dennis commented that the Team Wendy guy was right. "It's too soft. What he was not right about is what this can do when you do this [impact] very fast." Mike demonstrated by slamming his fist into a pad sitting on a table. He has broken glass tabletops in the past demonstrating the impact on his pad. He broke two antique tables in congressmen's offices. Yet he did not harm his hand, arm, or shoulder doing that because of the foam's molecular structure. The force of the hit to the foam and its resultant acceleration causes a molecular deformation, a controlled deformation, molecule by molecule, of the material that locks up during acceleration. "It doesn't want to be accelerated," Mike explained, "and for a millisecond becomes a solid. Everyone thinks it is pneumatic, but it is not. It is some serious physics happening here."

The Dan Rather piece was a big disappointment to Doc Bob. He felt it was a manipulation of the whole story and was weak, a "he said, he said" piece with no conclusions. Also, it did not "stress enough that troops in combat are the ones complaining about too-firm helmet pads and with good reason." In the end, the issue of comfort played a serious role for those in combat. "I see my Marines taking their hands off their weapons and playing with their kevlars all the time, and the first chance they get to pop tops, they take it," a Marine Corps Lieutenant wrote Doc Bob in August 2015. "The bottom line is the people who make these decisions don't wear our gear, and don't patrol and stand post in 100 plus degree weather so they write off the complaints as standard bitching. . . . They cannot quantify the argument that discomfort equals lax security, degrades endurance and speed with respect to the enemy."[24]

More recently a Marine Corps sniper complained to Doc Bob in September 2017 about supply-issued pads: "The pads turn to stone in the cold, lose any springiness once you sweat in them, and just can't fit your head properly whatsoever." He wanted more-comfortable pads to avoid headaches while sitting in a "hide" for seventy-two hours.

CHAPTER 26

Intransigent

Current pads are made by MSA. The main problem we are seeing is that the padding wears out or loses its shock absorbing/weight bearing properties quickly. As EOD techs, we need to be more concerned with addressing the IED threat outside the wire, not how our helmet padding is bothering us or affecting the fit of our helmet.

—U.S. Army first lieutenant in Iraq

"We were issued the Wendy pads which cause considerable pain, headaches and discomfort after wearing them. I'm interested in emailing PEO Soldier with my team's comments so they hear our feedback and make some changes to get us the best gear possible that is both functional and comfortable. I think these pads will reduce the number of complaints I receive from team members who have difficulty with their helmet pads."

—U.S. Army captain in Iraq

While Doc Bob was receiving these complaints and many more from Army soldiers in Iraq, PEO Soldier's intransigence continued. Not surprisingly it again was in a state of denial, insisting it had not received one complaint about the helmet-pad system from soldiers in the field.[1]

For an outsider like Doc Bob, fighting the military bureaucracy was like banging your head against the wall without a helmet. He was fighting an entrenched, obstinate bureaucracy that was more concerned about self-preservation than the welfare of the troops and disdainful of products and technology invented by "outsiders" that would upstage decisions within the bureaucracy. His efforts did raise the issue of padded suspension systems in combat helmets to a high level of concern, leading the Marine Corps, kicking and screaming, to padded systems in its helmets that may have saved thousands of marines from TBI or fatal brain injuries.

However, the battle continued between Oregon Aero and the helmet bureaucracy of both the Army and the Marine Corps. Oregon Aero was told in a 2009 meeting with PEO Soldier's chief scientist that all the helmet pads fielded by the Army in the past, including Oregon Aero pads, were qualified. Oregon Aero wanted an official announcement made by the Army that its pads were qualified. Instead, the company was assured that such information was disseminated to those who needed it, that anyone in procurement requesting the information could access it, which meant nothing unless the information was given to procurement directly as a policy.

Despite this assurance Oregon Aero's presumed qualification apparently was not making it to those who could act on it. To Oregon Aero officials, it sounded more like a whisper campaign was making the rounds to spread the opposite message. In the end nothing changed. The Army continued to say that Team Wendy pads were the only approved product.

In July 2008 the Marine Corps believed it could eliminate most of Doc Bob's concerns about its use of supply-issued helmet pads for its LWH. To accomplish this the Corps invited Doc Bob to a meeting at the its headquarters in Quantico, Virginia. Doc Bob brought his son, Mark, to the meeting, along with a congressional aide from the House Armed Services Committee who was investigating the helmet-pad issue. The Marine Corps contingent included high-ranking members of its equipment bureaucracy.

With the Marine Corps officials facing down Doc Bob and his son, the tension in the room was palpable. The officials immediately laid down their marker by telling Doc Bob that their supply-issued pads were the best for their marines. To Mark Meaders it seemed like the discussion was over at that point, given the officials' "we don't want to hear your stuff" attitude. The Marine representatives were steadfast in their belief in the USAARL tests that had disqualified Oregon Aero pads despite the fact that Congress had disputed them. They pointed out that Team Wendy pads tested better than other pads for temperature extremes.

Doc Bob countered that the temperature-extreme testing conditions—cooling the product to 14 degrees Fahrenheit and heating it to 135 degrees Fahrenheit—were not compatible with the survival of a marine. He tried to get across to the Marine Corps representatives that in the real world pads would seek a steady-state temperature somewhere between that of the scalp and ambient conditions. Using those figures in an "all temperature" table skews the results in unpredictable ways. Ambient or steady-state state temperature testing is more reliable. If ambient figures were used, the Oregon Aero pads would be slightly more protective than the Team Wendy pads. One of the Marine officials could only respond, in bureaucratic speak, that specifications for temperature testing were set in concrete, that adjusting test specifications for "real temperature" was not allowed.

Doc Bob also brought up the wearability problem with supply-issued pads as an important issue to be addressed. The Marine Corps's response made it clear that its top priority was protection over comfort for its helmet pads, even though a helmet that is removed because of discomfort cannot provide protection. To Doc Bob, however, if the blast/impact protection of pads is essentially equal under realistic conditions, wearability should be an important determining factor. A helmet that is removed because of discomfort is a dangerous helmet.

Doc Bob and his son, Mark, felt they had an amicable discussion with the Marines, meaning not much came of it. They also took away from the meeting the impression that the Marines were not interested in their opinion. Mark felt that the Marines did not want them in their business and wished they would just go away.

After the disappointing but not surprisingly wasted meeting with the Marine Corps, a determined Doc Bob decided to try his hand at meeting with Army helmets officials. In an email sent to officials at PEO Soldier requesting a meeting, his stated justification was the need to "reassure concerned families and troops that the military [was] taking steps to make their equipment both wearable as well as protective and not dangerously flammable." He pointed out the potential dangers of flammability and the too-firm pads that research scientists had determined enhance transmission of the blast wave from the Kevlar shell directly to the head. He wanted to finally resolve these issues and "be able to shut [his] doors and go play golf."

PEO Soldier officials hemmed and hawed about a meeting for a period of time, but Doc Bob persisted until he was able to arrange the meeting. He told them they had a problem, the troops were telling him it was a problem, and they needed to discuss it. They finally agreed to meet him in Washington DC, in the House Armed Services Committee's hearing room. On December 9, 2008, Doc Bob, along with a staff member from the House Armed Services Committee, met with the PEO Soldier's helmet officials to discuss the same issues addressed with the Marine Corps. Typical of the military's tendency to stack the deck, the PEO Soldier brought a contingent of twelve or thirteen people to the meeting. Strangely, one of the PEO Soldier officials actually passed out little goodie bags full of stuff made in China, such as tiny lasers, lip balm, and other frivolous items, similar to those distributed at trade shows.

Doc Bob's opening salvo included the fact that he had no financial relationship with Oregon Aero. He was there for the troops and didn't make a dime out of it. PEO Soldier officials responded by claiming they knew that he was not making money and had never said he had. There was no argument from PEO Soldier that the infamous USAARL tests, purportedly using Oregon Aero pads that failed extreme hot testing, were fatally flawed. Unlike the Marine Corps officials, its representatives admitted that the USAARL tests were not acceptable, that it was not what Congress intended, but the DOD had subsequently given them money to do testing at three civilian laboratories. They revealed that civilian laboratory testing was under way at the time of the meeting and would be reported in January 2009. However, they did acknowledge that preliminary information from those tests revealed that all pads passed 150 g levels. There was no statistical difference between them. They did not say that any pads performed worse than Team Wendy or that any did better.

Doc Bob would later request the results of the tests through the Freedom of Information Act. All he received from PEO Soldier was

185 pages of blacked-out test data, all information redacted for reasons known only to PEO Soldier. Doc Bob never learned the results.

Despite PEO Soldier's admission that the tests were flawed, Doc Bob asked the military officials about the differences in thicknesses of the pads used for testing that was reported in the infamous technical memorandum. The response was they hadn't noticed it. One of the PEO Soldier engineers attending the meeting admitted it wasn't honest to use the 2003 tests for the 2006 congressional report without labeling it.

PEO Soldier officials were very careful about what they revealed during the meeting. Their only answer to the blast-injury revelations made by outside researchers was that the effect of blast as a cause of TBI was still unknown. They claimed they were doing shock tube tests, placing a tube with a diaphragm on one end in water and measuring the transmitted force of a blast wave. When Doc Bob asked for the results, they claimed that the results were classified.

When asked by Doc Bob why their procurement decisions were based on the 2006 USAARL test-tests, they admitted those decisions were flawed and passed the buck to the NIB as the decision maker on which pads would be used. Doc Bob countered, "NIB has no scientific validity whatsoever." One of the officials then responded that PEO Soldier actually made the decision and gave the NIB the list of approved pads to choose from. When Doc Bob requested the list of approved pads, the officials suddenly changed their story, claiming they didn't have a list of approved pads. In fact they couldn't explain how the NIB had such a list.

Much as he had after the meeting with the Marine Corps, Doc Bob felt that PEO Soldier officials gave them the slow roll with nothing accomplished. He recalled a story had he heard when he was working in Africa, one he thought was apropos for how the meeting went. A Masai warrior described to him how to kill a goat without alarming it: "If you fiercely strangle a goat, the meat is stringy and tough as hell. It's different if you lay it down on the ground and stroke it very gently and turn its head a little bit. You stroke it, talk to it, sing to it, a little Masai song, turn its head a little bit more. Before it knows it, you have turned its head enough so you cut its windpipe, and it just goes to sleep and dies and the meat is tender. It doesn't have a bowel movement and defile the cooking area." That's what PEO Soldier was trying to do to him, Doc Bob thought—just stroking him and turning his neck slowly.

Although PEO Soldier officials had assured them during the meeting that soldiers were "authorized to use Oregon Aero pads donated by Operation Helmet," the Army issued a directive four months later reversing that assurance: "It has been reported that soldiers and units are purchasing individual soldier ballistic protection items, body armor, and ACH helmet pads through a variety of sources." It then directed that the "only authorized supply sources for ballistic protection items," including helmet pads, was "the Department of Defense supply system."[2] Afterward, thousands of soldiers ignored the directive and along with many marines continued to request Oregon Aero pad kits through Operation Helmet.

Three years later Doc Bob and his son, Mark, again met with PEO Soldier officials in Arlington, Virginia. True to the way DOD agencies work, different personnel had now been assigned to represent PEO Soldier.

Prior to this meeting, Doc Bob and Mark met with the helmet program manager. He was a hard-core infantry paratrooper with combat experience who had worn the PASGT helmet, the MICH, and the ACH with both Oregon Aero and Team Wendy pads. He noted the discomfort of the Team Wendy pads while at the same time praising them as a great improvement over the old padless PASGT; he had tried the Oregon Aero pads and liked their comfort. Doc Bob gave him copies of emails from both Army soldiers and marines regarding their reactions to the supply-issued helmet pads.

During the meeting the program manager emphasized that soldier comfort was now "a strong factor" in helmet-pad selection decisions and that all vendors would be treated equally and fairly. Doc Bob opined he was definitely a lone wolf, surrounded by career bureaucrats justifying their existence by holding on to the status quo. For example, the chief scientist for PEO Soldier, who also attended the meeting, again defended the testing of helmet pads at extreme temperatures, arguing that "some standard for comparison" was needed.

He did not seem to feel that changing the testing standards was worth the effort.

After the meeting two PEO Soldier officials approached Doc Bob and Mark, thanking them and saying that the fresh look at helmet pads and new technology would not have taken place without Operation Helmet. Doc Bob thought, it had only taken seven years to get someone to take a look, even if it was for naught.

CHAPTER 27

Too Little, Too Late

If you don't have blood coming out of your head, if you don't have a penetrating injury, you're fine, everything's okay, have a nice day.

—Dr. Christian Macedonia, military doctor and leader of the military's Gray Team in Afghanistan recalling military dictum in 2009

Old-school military thinking recognized only physical injuries sustained during combat as legitimate. Closed-brain injury and its subsequent long-term psychological effects were largely ignored. Unfortunately, the long-term psychological damage of TBI during the Iraq and Afghanistan conflicts resulted in a dramatic rise in suicides. In 2016, the *San Antonio Express* reported Pentagon statistics from all service branches totaled 4,839 suicides for the years 2003 through 2015 compared to 4,496 Americans killed while serving in Iraq.[1]

Historically, the combat helmet was never designed to counter the effects of blast and, until recently, blunt impact. The steel helmets of World War I and II were designed to protect the soldier from shrapnel and spent or glancing bullets. The Kevlar helmet was designed with primarily increased protection from penetrating ballistics in mind. It wasn't until the development of the ACH that blunt-impact mitigation was considered. But no requirement for blast-impact mitigation was developed despite the fact that since 2003, the vast majority of injuries during the wars in Iraq and Afghanistan were blast-related TBI. The reason was that the effects of exposure to blast were generally unknown and little understood.

Research on the effects of blast injuries on combat troops was rare during the last century. The first such published research occurred during and after World War I concerned the psychological aftereffects of closed head injuries, called "shell shock" at the time, which now are recognized as the result of exposure to blast injuries. However, no military medical literature reports of injuries in World War I attributed them specifically to blast exposure. British Army physicians Fred Mott, Gordon Holmes, and Charles Myers documented some of the first cases of blast injuries that occurred during World War I. The subjects of this research included survivors with no external injuries who nonetheless suffered severe headaches, amnesia, tremors, and other similar symptoms. Unfortunately, some soldiers during World War I who suffered from "shell shock" and other symptoms of blast exposure, such as what is now known as PTSD and TBI, were executed after being convicted of offenses such as cowardice and desertion.

Following the war in 1922, the British government, facing an enormous case load of injured soldiers on military pensions, produced a report dismissing any link between combat injuries and exposure to blast and shell shock, thus eliminating any discussion of that link for the next twenty years.

During World War II research on blast injuries increased. Even though it was known that blast could induce TBI, the causes were not solidly established nor were its effects understood. Even though research picked up in the 1990s and into the early twenty-first century, none was able to connect blast-wave trauma with measureable brain damage.

International researchers, including Aris Makris, at Med-Eng Systems, Inc. (a Canadian based research, design, and manufacturer of personal protective equipment, generally for worldwide explosive ordnance disposal operations, such as demining, the process of removing land mines), conducted in-depth research into protective equipment, including head protection, to protect deminers from the effects of blast. This effort began in earnest in the late 1990s, long before the military considered such protection.

In 2003 Med-Eng Systems reported that blast tests determined that the overpressure wave from a blast induced violent levels of acceleration on the head of the victim, which in turn can cause a range of concussive injuries. The researchers found that since the circumference of the helmet is greater than the head it is protecting, the helmet acts as a trap for the incoming blast winds, which can result in the head being accelerated at a greater rate than an unprotected head. At the time this finding significantly affected wearers of the padless suspension system of the PASGT helmet, which was used by troops in the early days of the Iraq and Afghanistan conflicts.[2]

Med-Eng researchers determined that both the head and the face needed protection from the devastating effects of a blast wave. The face needed protection from overpressure acting on the nose and ears that can induce blast-wave acceleration to the head. In 1999 Med-Eng Systems, in conjunction with the University of Virginia and the U.S. Army, published the results of their study in a report titled "Reduction of Blast Induced Head Acceleration in the Field of Anti-Personnel Mine Clearance." According to the report "there was a high probability for a fatal head concussive injury when a military helmet is worn without a visor, or when no head protection is worn. Properly designed helmet systems, which included a full-faced visor mounted on stable helmet platforms, were demonstrated to provide significant protection against blast-induced head acceleration." They also determined that the combination of the helmet and the face visor reduced head acceleration by a factor of 75–90 percent.[3]

In 2005 Doc Bob emailed Aris Makris about the ability of the helmet with a padded suspension system to protect a soldier from the effects of blast waves on the brain. Makris explained in his reply that a soldier wearing the PASGT helmet, without a padded suspension system, was vulnerable to blast due to the blast wave "being trapped and pulling the head in the direction of the blast, thereby inducing a blast-induced acceleration to the head."[4]

Makris further explained that the

blast wave itself reaching the head can possibly transmit across the composite shell, say if the exposure was from the side and rear, and transmit to the head. However, when the blast goes across the composite shell and finds an air gap, it is generally a positive thing so long as that gap can be preserved. It obviously is difficult to preserve in all cases, as evidenced by the helmet colliding with the head and the suspension system shifting over the head. Foam pads can take care of this issue and will generally be a positive development in maintaining a spacing between head and shell, as well as an air gap between the shell and head where foam does not reside.

However, Makris had learned from his research that a full face or partial face visor was more effective than the foam pads themselves in protecting the face and deflecting, or minimizing, the blast wave from penetrating the helmet cavity. He advised that "one has to look at the problem of blast protection holistically. Everything can help but one countermeasure alone may not cut it."

Echoing Makris's research finding that a full or partial face visor was more effective than the foam pads themselves, a helmet simulation study in 2010 by researchers at MIT also found that a standard ACH, modeled from helmet and pad material properties provided by Natick Soldier Research Development and Engineering Center, only slightly delayed the impact of the blast wave "but did not impede direct transmission of stress waves into the intracranial cavity because it does not protect the face . . . whereas the mitigating effect of the helmet-face shield combination is much more pronounced."[5] In fact according to Paul Scharre and Lauren Fish, researchers with the Center for a New American Security, "computer models have shown that improved helmet designs, such as with full-face shields can reduce the amount of pressure transmitted to the brain by up to 80 percent." However, there is a tradeoff in that a full-face helmet would be a little heavier and limit visibility.[6]

Defense expert Pierre Sprey attributed the high frequency of blast-wave injuries in Iraq and Afghanistan to the preponderance of IEDS, which injured soldiers more often than rifle bullets did, something the Army knew by 2004. By then military hospitals were filling up with blast-related injuries. According to Sprey most of those injuries involved soldiers in enclosed vehicles, where the blast effect is magnified by a factor of three. Those injuries would not have been as severe in the World War II jeep because it was open and thus no magnification of the blast would have occurred. The helmets then were more open, with no coverage at the back of the neck, and didn't magnify the blast to the skull like the current helmets did. The newer helmets covered more area to prevent more fragment injuries, which was good, but the design increased the blast injuries.

"Iraq has brought back one of the worst afflictions of World War I trench warfare: shell shock," the *Washington Post* reported. "The lethal blast wave is a two-part assault that rattles the brain against the skull. The initial shock wave of very high pressure is followed closely by the 'secondary wind': a huge volume of displaced air flooding back into the area, again under high pressure. No helmet or armor can defend against such a massive wave front."[7]

Concerned about the rise in TBI after exposure to roadside IEDs, researchers at the Army's Fort Carson, in Colorado Springs, Colorado, conducted a twenty-two-month study of its soldiers returning from Iraq. The study of about 13,000 soldiers, examined from 2005 to 2007, found that 2,392 of them exhibited some of the symptoms associated with mild TBI. Fort Carson Army medical officials felt that the effects of blast exposure explosions that "rattled the head of Fort Carson soldiers" were "overwhelmingly to blame."[8]

Additionally, the previously mentioned RAND Corporation study released in 2008 with data collected between April 2007 and January 2008 disclosed that as many as 320,000 service members returned from Iraq and Afghanistan with symptoms of TBI from exposure to blast. The RAND study referred to the alarming rise in reported TBI cases as "the signature wounds of the Afghanistan and Iraq conflicts."[9]

Despite international research conducted on the effects of blast on helmets, mainly for demining operations, by 2006 the military had still seemingly made no effort to conduct similar research. During the congressional hearings on helmet pads in June 2006, Marine Corps general Catto admitted to committee members that the Marine Corps did not have the money to do blast research. This admission was made despite the report published just a month earlier, in May 2006, by the Marine Corps Center for Lessons Learned recommending that protective equipment such as helmets "should be evaluated for potential design modifications to ameliorate or prevent TBI from blast injury."[10]

It appeared that the military had failed to connect the dots between a huge rise in closed head injuries from blast exposure and the design of the combat helmet—the only piece of equipment the soldier or marine had to protect the head. Also, given its helmet-pad testing criteria, the helmet bureaucracy seemed not to understand the difference between localized brain damage from blunt impact and diffuse brain damage from blast impact. According to Dr. John D. Lloyd, a board-certified ergonomist, brain injury specialist, and consultant in biomechanics, blunt impact produces a linear or localized head injury while blast impact produces a rotational or diffuse effect on the brain. "Typical helmet tests only measure forces associated with linear acceleration and therefore fail to account for the risk of brain injury," Lloyd maintains. Thus, new criteria needed to be established to test the helmet and pads that induce "rotational inertia on impact, thereby facilitating measurement of risk of focal and diffuse brain injuries."[11] The helmet bureaucracy's delay of important blast-injury research led to tragic consequences for the troops during the early years of the war in Iraq and Afghanistan. Hit with a huge rise in TBI caused largely by exposure to blast, the military equipment bureaucrats, inexplicably caught with their pants down, had no playbook to deal with it. It would literally take an act of Congress to get the military to implement a program to research blast-related injuries.

Finally in 2006, concerned over the high incidence of troops experiencing TBI from exposure to blast, Congress acted to force the military to start conducting blast research. It mandated, through the fiscal year 2006 National Defense Authorization Act, that the DOD establish a blast-injury research program. The program was established on paper in July 2006, with the secretary of the Army delegated as program authority.[12]

Due to a slow-moving bureaucracy, almost an entire year passed before the Blast Injury Research Program was implemented. In June 2007, four years into the Iraq war and after blast-related injuries had become common, involving thousands of soldiers and marines, the research program finally got under way. Its stated purpose was to "coordinate and manage relevant DOD research efforts and programs for the prevention, mitigation, and treatment of blast injuries."[13]

Like so many agencies within the DOD, this new program created a bureaucratic empire, composed of a dizzying array of many existing and new acronym-laden agencies that together formed an esoteric structure while performing any number of functions. The program was coordinated under the Blast Injury Program Coordinating Office. Managers from each of the military departments along with science advisors and networks for incident analysis and injury prevention, among others, were gathered under an agency called the Joint Trauma Analysis and Prevention of Injury in Combat Program (JTAPIC). An executive authority (also called the executive agent) for the program was assigned to the secretary of the Army, but later that responsibility was delegated to the commander of the Army Medical Command.

Shortly after the Blast Injury Research Program began in 2007, the Army's vice chief of staff directed JTAPIC to develop helmet sensors to measure acceleration and overpressure forces on the helmet. Called the Gen I Helmet-Mounted Sensor System (HMSS), the sensors were installed on the helmets of two Army brigade combat teams (BCTs) and two Marine battalions during their deployment in Iraq between December 2007 and February 2008.

The objective of the sensor program was to collect data from the sensors that would hopefully provide a better understanding of the nature and frequency of head trauma loads facing the troops in combat. The sensors only measured exposure to blast—helmet acceleration and pressure from impact. They did not indicate injury. One goal of the program was to acquire a better understanding of those forces leading to the development of the next generation of combat helmets. Unfortunately, the results from the first-generation sensor program were minimal, primarily because of sensor problems. The sensors did not produce accurate recordings. Date and time stamping, correlating the event with the time, was a problem. Blast-wave angles also confused the sensors. The bottom line was that Gen I was not a reliable test mechanism to capture blast data.

After going back to the drawing board, the designers produced a second-generation helmet sensor program, named Gen II HMSS. This sensor was designed to capture linear and rotational acceleration and blast pressure. Forty-five thousand Gen II helmet sensors were fielded between November 2011 and February 2012. However, the program's Report to Executive Agent for fiscal year 2012 revealed that evaluations of the collected Gen II sensor data, as with the first generation, "failed to provide readings that correlated with the effects of the blast." The development of yet another new generation sensor system brought hope of ultimately correlating "all combat-related traumatic events to actual head injury for Soldiers."[14]

Along with the Gen II sensor, the Defense Advanced Research Projects Agency (DARPA) developed a blast dosimeter, the DARPA Blast Gauge, that was fielded to three combat units. The blast gauge also measured blast exposure while rating the level of exposure and recording and storing exposure data for later research. About the size of a quarter, the blast gauge could be attached to the helmet or on other gear worn by the warfighter.

Yet by 2014 very little had been accomplished with the use of the sensors in determining injury mechanisms resulting in TBI from either blunt- or blast-impact exposure on the battlefield. It was reported at the November 14 International State-of-the-Science meeting that of 378 events of impact exposure to those in combat wearing helmets sensors, the sensors triggered a warning in only 12 cases; the sensors missed 93.1 percent of concussions among the potentially concussive events.[15]

In late 2016 National Public Radio's (NPR) *Now*, the media organization's twenty-four-hour program stream, reported that the Pentagon had quietly discontinued its DARPA Blast Gauge program because it wasn't measuring enough exposures from IED explosions; "the gauges failed to reliably show whether service members had been close enough to an explosion to have sustained a concussion or mild traumatic brain injury." The secretary of the Army, in a letter to Representative Louise Slaughter (D-NY), explained that the gauges were unreliable in providing "data in the training or combat environment."[16]

However, as an unintended consequence, while the gauges were not reliably measuring blast exposure from IEDs, they were, instead, recording data of blast exposure from troops firing or even being near heavy weapons such as recoilless rifles, shoulder-fired rockets, artillery, and mortars. In other words soldiers firing their own heavy blast-intensive weapons were also taking a jolt to their brains from a pressure wave generated from firing them.

Shelving the gauges resulted in criticism from military veterans including retired Army vice chief of staff Peter Chiarelli, who told NPR that the shelving was "a huge mistake." He believed that the sensors could play "a very, very important role in helping [the military] understand why an individual ha[d] negative effects from a concussion."[17] "Here's something you learn in data science: Sometimes the answer you get was not the answer you were looking for," wrote Elana Duffy in an article on *Task & Purpose*. "But as long as the data isn't corrupt, even if it isn't exactly the data you were seeking, you can still use it. When you start research, you should be open to getting results you weren't expecting, and tweaking the method if you need different information instead of stopping the study. Instead, DoD just shot the messenger. They didn't get results from enemy weapons, they got it from our own (and a little of the enemy), so they shelved the technology."[18]

After two years of the Blast Injury Research Program, with its Byzantine bureaucratic structure, multiple research efforts, sensor program, and thousands of additional soldiers afflicted with blast-induced TBI, very little progress had been made toward achieving the program's mission. Not much was accomplished that led to better head protection for the troops. Instead, the result of two years of work produced more questions than answers, with no clear indication that blast-induced mild TBI even existed as an injury. "It is not known whether non-impact, blast-induced mTBI exists as a unique injury," Michael J. Leggieri Jr., director of the Blast Injury Research Program Coordinating Office, stated at a Biomedical Science & Engineering Conference in 2009. "If it does exist, the mechanism of injury is unknown. Despite this lack of knowledge, many people mistakenly expound or assume that exposure to blast overpressure alone can cause mTBI."[19] This sounded much like a page out of the NFL book on football-related concussions and TBI.

By 2010 it still appeared that no progress had been made in identifying mechanisms of brain injury due to blast exposure. "There was insufficient evidence to support one mechanism as the most plausible explanation for non-impact blast-induced mTBI," Leggieri explained. "There are insufficient data on the nature of non-impact, blast-induced MTBI to make recommendations on how to better protect Soldiers." Therefore, according to the Leggieri, "[We] can't protect against an injury that we don't fully understand. Until we see conclusive data supporting the existence and mechanisms of this particular injury, there should be no attempts to modify existing protective equipment such as combat helmets."[20] MIT researchers echoed this problem in 2011, stating that from their studies "little [was] known about the mechanical effects of blasts on the human head, and still less [was] known about how personal protective equipment affects the brain's response to blasts."[21]

In an April 2011 *Defense.gov News–American Forces Press Service* article, Leggieri said, "Impact is something we know quite a bit about. But this whole question about blast is still a question. And although the Army is at work on its second-generation helmet sensor with plans to field it soon to about 30,000 soldiers, there's still no clear indication of what those blast readings will mean in terms of the brain."[22]

As late as September 2013, six years after the start of the program and despite the many agencies and individuals working on blast-induced injury, there was no apparent progress determining the mechanisms of injury and thus a new helmet design to protect the troops from TBI. The program's annual report for fiscal year 2012 stated, again, "The current understanding of the existence and mechanisms of non-impact, blast-induced MTBI is very limited. There are numerous hypotheses of the mechanisms of brain injury by blast exposure to the head."[23]

This result was again echoed in 2014, when David Mott, a Naval Research Laboratory aerospace engineer, told the *Army Times*, "The military actually has specific criteria that helmets have to meet to be certified for use in ballistic and blunt force. No such criteria exists for pressure because the medical community is still working on what the injury mechanisms are, and we don't know where to set those desirable levels."[24]

The large number of people and agencies that make up the Blast Injury Research Program bureaucracy may have contributed to its inability to identify a mechanism. This bureaucratic structure by committee, with its varied theories of what happens to the brain when exposed to blast and conflicting information that could have nullified a consensus, may forever prevent the program from making decisions that would lead to designing better helmet protection. Pierre Sprey felt that despite Congress mandating a blast program, the program had done nothing, and "the Army doesn't even know how to write a specification for blast."[25] While the DOD's Blast Injury Research Program was trying to identify the mechanisms leading to TBI from exposure to blast in order to design a combat helmet to mitigate or eliminate such injuries, research was being conducted independently around the country and internationally on identifying the extent of damage to the brain along with injury mechanisms.

On a Sunday morning in 2008, William Moss, a researcher at the Lawrence Livermore National Laboratory in Livermore, California, was having breakfast with his wife, a PhD in neuroanatomy. She was reading an article in the *San Jose Mercury News* about the hidden wounds of the Iraq war and closed head injuries. She showed the article to her husband and asked him if he could simulate the circumstances causing such injuries. He thought he could. The next day he approached his colleague Michael King, and with a little program money, they started a research project to address the general question of what a blast does to the skull. They were not thinking about a helmet or mechanism but just blast-skull interaction and if any significant factors came into play.[26]

Working on their own, they conducted their study with simulated helmet models, one with a padded suspension system found in the ACH, and another with a web suspension system found in the PASGT. The explosive, a fairly standard five pounds of C4, was located fifteen feet from the helmet models. They discovered that blast and impact were very different. They noticed immediately a skull flexure mechanism, or skull flexural ripples, caused by the blast sweeping over the skull, much like a rolling pin on dough creates ripples ahead of it or a train going over tracks sends ripples ahead of it. They did not know whether the flexure caused TBI. They just knew that it created the effect. The pads they studied were not representative of any commercial pad. They were pads representative of the material properties they thought were reasonable.

The primary effects of blast on the skull should be nonexistent because the skull is hard and the blast is just wind. Moss and King were dumbfounded that they could get that effect with the flexure mechanism. They are not aware of it being identified in previous studies.

They didn't have enough detail in the simulations to know if the structure of the brain was damaged. All that they could say was that the wave created a region of high pressure and one of low pressure that moved around. The loads were about the same order of magnitude, and they were about as large as the loads seen in the brain in injuries caused by impact, but the patterns of those loads were very different at impact. At that point Moss and King could not necessarily say that the loads caused injury, but it didn't seem too far a stretch of the imagination to think that it might.

With the web suspension system helmet in the simulation, the pressure wave could be seen washing between the helmet and the skull. There was a focusing effect with the open web system. The pressure between the helmet and the skull was greater than the blast wave outside. With an open web suspension system, the gap between the skull and the helmet allowed the pressure wave to wash up into the gap causing a focusing effect that made the pressure worse. With a padded suspension system, Moss claimed that the pad prevented the underwash effect. Also, by coupling the helmet shell to the head with pads, the loads on the brain were strongly dependent on the pad's material properties.

Importantly, Moss noted, like Aris Makris in his research, that a foam structure with increased stiffness transferred loads more effectively. Moss and King then tried to figure out if it was better to have higher stiffness initially and then a low response or a low stiffness initially and a high response. They conducted a number of tests with the Oregon Aero pad to get what they thought was the right order of magnitude of average density, then increased it three orders of magnitude, decreasing the magnitude at certain intervals. As a result Moss and King were able to pick points of stiffness that had lower loads and points that had higher loads. What they learned was that the loads got worse as the magnitude increased.

Moss and King found that the cellular characteristics of the foam were different depending on whether it was subjected to blunt impact or blast. The loading and the rate would be different, and the total amount of deformation would be different. After testing the baseline pad model, they saw it was better than the web suspension system and a little better than no helmet at all. They learned that an increase in foam stiffness increased the coupling of the external loads to the head. In hindsight Moss felt it was almost common sense because as the medium transmitted the threat, the more stress was transmitted. It's like putting a marshmallow on your head and pushing on it, which would be fairly easy, but if you put a steel block on your head and push on that, it's going to be harder to push than the marshmallow. Moss and King concluded that loads transmitted to the brain in their simulations were larger if the pads were stiff rather than soft.

Following their results Moss and King briefed various agencies in the DOD about their work, partly to find out if they were crazy and partly to tell them what they had done to determine if there was any interest in follow-up funding. The Blast Injury Research Program Coordinating Office director, Michael Leggieri, was one of the officials they spoke to, along with the Army medical commander, explaining the differences between blast and impact. According to Moss and King, they could see that many in the room knew, in their gut, that blast and impact were fundamentally different. The two researchers, however, were the first to quantify exactly what the differences were, based on their simulations of blast and impact. The detailed computations were a little bit outside the realm of what, at the time, the Army was used to seeing, but the researchers realized they were not dealing with ignorant people. Moss and King also briefed House Armed Services Committee staffers and various congressmen in an attempt to generate interest in funding follow-up work.

In September 2009 Moss and King's professional research paper was published, detailing their findings about skull flexure from blast waves and its implications for helmet design. They pointed out that the skull flexure mechanism has "implications for the diagnosis of TBI in soldiers and the design of protective equipment such as helmets." Discussing ACH-style helmets with a supply-issued foam padded suspension system, they noted that "underwash is mostly prevented but the helmet is more strongly coupled to the head, so helmet motions (bulk acceleration) and helmet bending deformations are transferred directly to the skull."[27]

Since Moss and King had determined that the stiffness of the foam was sensitive to rate dependency, they felt that "stiffer foams transferred greater loads from the helmet to the skull, reducing the helmet effectiveness against overpressure." Thus for helmet design "an effective mitigation strategy against the deleterious effects of skull flexure would be to deny the blast wave access to the airspace under the helmet, and then either incorporate rigidity into the helmet itself, or design the helmet suspension system so that the flexure of the helmet is not transferred to the skull."[28]

For Doc Bob Moss and King's findings seemed to validate his argument against the stiffer Team Wendy pads—not only for wearability concerns but also because stiffer pads would transmit blast waves to the skull more so than the softer pads. He believed the computational modeling were very sound and had accurately determined that too-firm pads make the skull flexion problem worse. Additionally, brain injury expert Dr. John Lloyd felt that stiffer pads could increase rotational inertia to the brain, leading to the risk of TBI.

As a result Doc Bob believed that Oregon Aero pads allowed expansion in all directions without bottoming out except at extremes, while the other two pad systems currently in use did so only to a limited degree. They all passed g-force drop tests in the lab but differed considerably in acceptance when in use in combat, according to the responses he received from troops.

Given this credible research, Doc Bob questioned how long it would take the military to actually utilize the fruits of national research institutions and protect the troops. TBI was costing U.S. taxpayers approximately $1.5 billion a year, he argued, not to mention the costs to the soldiers or marines and their families for the rest of their lives. Professor Eric Blackman of the University of Rochester, who assisted Moss and King in their research and coauthored the research paper, believed that the military generally ignored academic research and did not seem as driven to make TBI a high priority as it was to identify IEDs and protect against them.[29] It was estimated by various sources that as of 2011 the DOD had spent around $17 to $19 billion on contractor IED detection gadgets that only worked 50 percent of the time—a huge waste of money, some of which could have been spent on more expensive but wearable suspension system pads that would have helped troops to be more efficient in doing their jobs. In fact training dogs and their soldier handlers to detect IEDs cost only $8.7 million and had an 80 percent effective rate.[30]

Since Moss and King were outsiders and were not conducting DOD-sanctioned research, their work was dismissed outright within the Army's equipment bureaucracy and the Blast Injury Research Program. The Army bureaucracy appeared to put more effort into rejecting this research than into considering it as a credible mechanism for TBI. Again the "not invented here" mentality was at work. Dr. James Zheng, the chief scientist for the Army's Soldier Protection and Individual Equipment program, told Doc Bob that he did not believe the Moss and King findings regarding skull flexure or the stiffening effect of pads.

The Blast Injury Research Program's Michael Leggieri echoed this attitude in 2011, continuing to insist that there was insufficient conclusive data to support any one mechanism. He felt that models based on simulations couldn't be scientifically validated. He believed what was needed was a "validated mathematical model" to show how a blast interacts with the skull.[31]

Yet by the time the program's submitted its fiscal year 2015 Report to the Executive Agent, it recognized skull flexure as one of several mechanisms believed to contribute to blast-induced TBI. Still, the report claimed, "There is not yet a quantitative description of the specific contribution of each of these mechanisms."[32]

Following their initial research on blast effect on the brain, Moss and King wanted to extend their effort with a contract from the DOD. It wasn't long before they received a call from Col. R. Todd Dombroski, an osteopath and Joint Improvised Explosive Device Defeat Organization surgeon and medical officer. He was interested in comparative studies of NFL football helmet pad systems versus military pad systems. This interest surprised Moss and King. They had thought any DOD contract was going to be for a blast study and a follow-up to their groundbreaking research. Colonel Dombroski apparently had data indicating that a larger percentage of injuries evidenced some aspect of blunt-force impact than just pure blast, and he wanted to research impact.

The main thrust of the research was to lend some scientific insight to claims that football helmet pads might work as well or better than the pads the Army was currently using. Moss and King were given the general guidance to compare the Team Wendy pads to the Riddell football helmet pads used in the NFL. They also wanted to do the test with Oregon Aero and Xenith pads. In the end their research became a comparison of the Oregon Aero crown pad with the Team Wendy crown pad used in the ACH. They picked the crown pad because it was easier to test.

From their research they determined that pad thickness played a significant role. "That was spectacular," Moss said.[33] They found that adding a little more thickness to pads had a big effect on blunt-force impact mitigation. They determined that the three-quarter-inch pad was right in the sweet spot. Hypothetically, if the pads were already two inches thick, adding another one-eighth inch would probably not matter. They also determined that the Team Wendy pad was much stiffer than the Oregon Aero pad and thus provided more impact protection at certain velocities. This was not a surprise to Doc Bob since a harder pad material would naturally offer more protection. "A pad made of concrete would offer a lot of protection but would not be wearable," he added. Doc Bob was disappointed that wearability between the two pad systems was not addressed.[34]

However, Moss and King focused on the Team Wendy crown pad as the baseline for the thickness determination because it was the official pad being used in the ACH. They did not do a thickness test on Oregon Aero pads. Moss conceded that if they had tested Oregon Aero pads, it most likely would have made a difference because the softer Oregon Aero foam actually densified a little sooner than the Team Wendy foam. Moss felt that adding an extra one-eighth inch might have more impact with the Oregon Aero pads.

King said he wasn't surprised that making a pad thicker made it better. It showed the beneficial effect of having a slightly thicker pad. According to King it was up to the Army to do whatever it was going to do with their research. In a press release, the head of PEO Soldier, General Fuller, said that the study proved the ACH offered the most protection regardless of the specific nature of the study results. Moss and King felt that statement was too broad since they did not study the entire pad suspension system of the ACH.[35] Yet the Army seized on it, and it wasn't long before PEO Soldier announced new one-inch pads.

While debunking Moss and King's findings of skull flexure, the Army's Dr. Zheng also insisted to Doc Bob in a meeting in May 2011 that no one knows what causes blast-induced TBI without visible injury and evidence of impact. Simulations and computational models by many researchers showed that a blast wave impacting the head, even through a combat helmet, results in a concussion that stresses the delicate soft tissue of the brain, causing damage leading to TBI. However, what type of damage does the blast wave cause once it hits the brain? Finding out would take a combination of animal and human brain autopsies to determine actual damage. By the time Dr. Zheng told Doc Bob that no one knows of the damage caused without evidence, a transformation in methodology was taking place, resulting in evidence of visible injury discovered through brain autopsy studies of deceased military veterans, primarily by independent organizations.

One study determined that concussions caused by exposure to blast appear to be different from concussions from blunt-force impact. In 2009 scientists from the Defense and Veterans Brain Injury Center studied veterans with diagnosed TBI from exposure to blast and those with TBI caused blunt blunt-force impact and found "those with blast-linked trauma had a more diffuse pattern of damage to the white matter, described as a 'pepper-spray pattern,' than those whose concussions were caused by direct impact or acceleration."[36]

New evidence was emerging that correlated brain damage from exposure to blast to brain damage from blunt-force trauma as experienced by athletes in football and hockey. Dr. Bennet Omalu, the chief medical examiner in San Joaquin County, California, who is associated with the Brain Injury Research Institute (BIRI), is an expert in forensic neuropathology. He was the first to discover a type of serious brain damage called chronic traumatic encephalopathy (CTE) in a former NFL player who had suffered repeated concussions and eventually committed suicide. A book depicting his discovery and later battle with NFL officials, *Concussion*, was published in 2015 and later became a movie starring Will Smith.

In early 2010 Dr. Omalu may have been the first to discover CTE in the brain of a deceased military veteran. CTE leads to progressive debilitating impairments such as behavioral changes, cognitive dysfunction, and dementia found in cases of TBI and PTSD, often leading to suicide. Dr. Omalu examined the brain of a sixty-one-year-old deceased Army veteran who served in Vietnam and was later diagnosed with PTSD. He found a buildup of harmful proteins in the brain consistent with CTE. "This is a sentinel case," Dr. Omalu said. "The brain findings in this deceased Army veteran are similar to the brain findings in deceased former football players. Now, we need to look at more brains."[37]

About a year later Dr. Omalu, along with colleagues at BIRI, identified CTE in the brain of a twenty-seven-year-old former Marine Corps Iraq war veteran (2006–10) diagnosed with PTSD who had committed suicide. It was determined to be similar to CTE found in athletes. The veteran experienced exposure to mortar and IED blasts less than fifty meters away. Dr. Omalu believed that the veteran's military exposure to "repeated subconcussive and concussive traumatic brain injuries" was the primary cause of CTE. He and his colleagues reported "this case as a sentinel case of CTE in an Iraqi war veteran diagnosed with PTSD to possibly stimulate new lines of thought and research in the possible pathoetiology and pathogenesis of PTSD in military veterans as part of the CTE spectrum of diseases."[38]

Dr. Omalu's findings on the West Coast were followed by similar findings on the East Coast. Dr. Ann McKee, a professor of neurology and pathology at Boston University School of Medicine, is also director of Neuropathology Service for the New England Veterans Administration Medical Centers. Like Dr. Omalu, in 2014 she discovered numerous cases of CTE in deceased former professional athletes. Dr. McKee and her colleagues examined the postmortem brains of four veterans of the Iraq and Afghanistan wars who were exposed to blasts on a repeated basis. They found evidence of CTE along with "myelinated fiber loss, axonal degeneration, microvascular degeneration, and neuroinflammation." The CTE was similar to the CTE found in former professional athletes. Medical histories of the deceased veterans showed that they were diagnosed with TBI along with PTSD. Based on her examination of the brains, Dr. McKee concluded that "repetitive mTBI is associated with CTE."[39]

Following the conclusion that blast exposure is associated with TBI and CTE, Dr. McKee and her colleagues looked at the primary injury mechanism leading to TBI and CTE. From research conducted on mice exposed to controlled blast, it disclosed a link between blast exposure and CTE. They reported: "It is notable that exposure to a single blast in our mouse model was sufficient to induce early CTE-like neuropathology." It was important to note from the results of their experiment with the brains of mice that the findings "point to the substantial inertial forces and oscillating acceleration-deceleration cycles imposed on the head by blast wind (bobblehead effect) as the primary biomechanical mechanism by which blast exposure initiates acute closed-head brain injury and sequelae, including CTE."[40]

Neuropathologist Dr. Daniel Perl is a professor at the Uniformed Services University and leads the Center for Neuroscience and Regenerative Medicine (CNRM) Brain Tissue Repository, established by the DOD in 2012 for TBI research. He is considered a leading researcher in CTE. In early 2012 Dr. Perl examined the brain of a former soldier who was exposed to a blast at a distance of five feet. The soldier survived but died two years later of a drug overdose. As in Omalu's and McKee's research, his examination of the soldier's brain disclosed scarring that was typical of CTE. Dr. Perl followed with an examination of several more brains of deceased military veterans with the same pattern of scarring.[41]

From the autopsies conducted on the brains of deceased veterans showing CTE, it is not known what type of helmet they were wearing while in combat or whether the helmets included a padded suspension system and, if so, what type of pads were included. Other research testing level of impact force on the head with various padded systems has shown a significant reduction in impact on the head with use of pads. How that translates to the degree of impact of a blast wave on the soft tissue of the brain and resultant injury, if at all, has not yet been determined. However, the information from the autopsies of postmortem brains could be used to design better helmets with upgraded padded suspension systems.

Despite these findings the Blast Injury Research Program reported in its fiscal year 2016 Annual Report to the Executive Agent that an International State-of-the-Science meeting, a forum of researchers in and outside the DOD, met in 2015 to discuss the relationship between head injury and CTE. They first determined that "existing research [did] not substantively inform whether the development of CTE is potentially associated with head injury frequency (e.g., single versus multiple exposures) or head injury type (e.g., impact, nonimpact, blast)." In addition, they concluded, "The current state of the science does not allow for a conclusive determination of whether exposure to head injury is associated with the development of CTE pathology or clinical symptoms."[42]

Later in the year, the expert panel in another International State-of-the-Science meeting concluded that, despite Dr. Omalu's and Dr. McKee's findings, "existing scientific evidence [was] insufficient to link blast-related TBI with CTE." Their recommendations included analyzing one hundred additional brains.[43]

In January 2016 a consensus panel of neuropathologists, sponsored by the National Institute of Health's National Institute of Neurological Disorders and Stroke and the National Institute of Biomedical Imaging and Bioengineering, met to define the neuropathological criteria for CTE. As a result of the Omalu and McKee studies, the group was in agreement regarding the diagnoses of CTE and the specific requirements for the diagnoses. They agreed that future investigation is needed to better understand the relationship of CTE to its biomarkers.[44]

In June 2016 scientists analyzing autopsied brains of former veterans with exposure to blast found evidence of distinct astroglial scarring (response to severe tissue damage or inflammation leading to scarring). The journal *Lancet Neurology* published the study to determine if exposure to blast caused damage to the brain in the form of brain lesions that induce TBI. The study by researchers affiliated with the DOD's CNRM, including Dr. Daniel Perl, analyzed postmortem brain specimens of eight veterans with both chronic and acute blast exposure. They determined that veterans with chronic blast exposure showed "prominent astroglial scarring," in the form of a "brownish dust," in certain regions of the brain, while cases of acute blast exposure showed "early astroglial scarring in the same brain regions." The study interpreted the findings as indicating "specific areas of damage from blast exposure." All the chronic blast exposure cases "had an antemortem diagnosis of post traumatic stress disorder."[45]

"We made what we believe is a significant breakthrough," said Dr. Perl. It may be a cause of the "physical and behavioral changes seen in some troops after they return from war." The scarring found in the brains differed from what is known as CTE. According to Perl, "CTE is an accumulation, and takes years to develop.... The service members we are describing develop these symptoms rather quickly, and they come back from deployment with these persistent symptoms. We believe it's related to the damage of the blast wave."[46]

Under the Veterans Administration, the Office of Health Equity adopted the findings of Dr. Omalu and Dr. McKee and is said to be leading the discussion by creating awareness of the impact of TBI and CTE. It envisioned that "best practices will emerge from VA on the issue of TBI and CTE." During an expert panel convened by the Office of Health Equity in June 2016, Dr. Omalu, as one of the presenters, said that CTE is part of a spectrum of diseases of TBI called traumatic encephalopathy syndrome (TES) that also includes post-traumatic encephalopathy (PTE). According to Dr. Omalu CTE is a progressive disease, meaning that it gets worse, while PTE is not progressive—once it is acquired, it does not get worse. PTE is described as the astroglial scarring described in the study by CNRM in the journal *Lancet Neurology*. In other words exposure to blast could manifest itself as either CTE or PTE.[47]

Currently, the only way to diagnose CTE is by examining a postmortem brain. However, scientists are hard at work to come up with a valid test to diagnose CTE from the symptoms of live patients in order to create a treatment plan. According to Dr. Ann McKee, scientists are closer than ever to devising such a test. McKee identified a certain protein found in brain tissue that could lead to such a test. It will require more extensive research to validate this finding, but she is optimistic about the results.[48]

However, a recently discovered blood test can identify two protein markers that appear in the blood following a head injury, leading to a diagnosis of TBI. Banyan Biomarkers, funded by the DOD and the Army, discovered the test and the equipment to administer it called the Banyan Biomarkers' Brain Trauma Indicator (BTI). The BTI was approved by the Food and Drug Administration on February 14, 2018, and has the potential for the military health system to identify TBI in a soldier following exposure to blast and getting that person the appropriate treatment.[49]

In January 2018, Dr. McKee, on a segment aired on *60 Minutes*, revealed that out of 102 veteran's brains she had examined, 66 had CTE, clear evidence that repeated exposure to blast can lead to this deadly degenerative disease of the brain. Dr. McKee explained that blast impact on the brain "causes a tremendous sort of . . . ricochet or . . . whiplash injury to the brain inside the skull and that's what gives rise to the same changes that we see in football players, as in military veterans."[50] Additionally, Luke Ryan reported in *SOFREP News* that "the blast wave does not only smack the head and cause the brain to strike the inside of the skull—it also passes *through* the head. It can do damage to a good portion of the brain this way."[51]

In addition to the discovery of CTE in veterans exposed to blast, later research also found brain scaring from exposure to blast in veterans diagnosed with PTSD. A CBS *60 Minutes* report on April 1, 2018, noted that this discovery by Dr. Perl "could mean that many cases of PTSD, long thought to be a mostly psychological illness, may actually be caused by physical brain trauma" from exposure to high explosive blasts. When the *60 Minutes* correspondent asked Dr. Perl if he believes a link can be established between blast-wave brain injury and PTSD, he responded: "In a sense, we already have. Every case that we have looked at has been diagnosed with PTSD."[52]

After ten years in existence, the Blast Injury Research Program, in its fiscal year 2016 Annual Report, stated that it still believed there was no consensus regarding the existence or the mechanisms of blast-induced TBI. Therefore, the program concluded that it was not possible to design a helmet to prevent or mitigate blast-induced TBI.[53] However, given the findings Dr. Omalu, Dr. McKee, and Dr. Perl indicating that a serious degenerative disease of the brain can develop from exposure to blast, tests of current helmets with padded suspension systems need to be conducted to determine if they have the capability to mitigate serious damage to brain cells that result in CTE.

CHAPTER 28

Back to the Future

B y the beginning of 2014, more than seventy-seven thousand helmet pad kits were donated to the troops in Iraq and Afghanistan by Operation Helmet. At the same time, the Army and the Marine Corps continued their claim of receiving no complaints about their helmets and pads while forbidding the purchase of pads from outside sources.

With the discovery that exposure to blast can lead to brain damage in the form of a serious degenerative disease called CTE, helmet design that absorbs the energy from blast overpressure impacting the brain became essential. According to Eric Blackman, professor of physics and astronomy at the University of Rochester in New York, "An immediate goal of helmet design should be to identify and require better impact protection cushioning and an important longer term goal should be to develop correlations between external and internal measures of TBI, where external refers to forces on the head form and internal refers to the resulting localized forces and stresses on the brain tissue."[1]

Impact testing for helmets and padding in the past has primarily focused on blunt-force impact, not specifically impact from blast. As shown in previous chapters, helmets with padded suspension systems have the ability to absorb much of the energy from blunt-force impact. To what extent the padding also absorbs impact energy from blast is not well known. Blackman believes that "internal cushioning that reduces impact injury may also reduce overpressure injury by damping the waves that would otherwise propagate into the head from skull flexure. This is an important direction for future research—finding the internal cushioning that optimizes the needed protection against both overpressure and impact injury."[2]

In 2007 the Marine Corps, led by Commandant General James Conway, decided to push for a new helmet with better ballistic protection. The design, production, and fielding of the helmet was given top priority. Unfortunately, there was no mention of designing the new helmet to better protect marines from blast impact despite rising cases of TBI among marines. It seemed to Oregon Aero's Mike Dennis that "everybody [was] lost in the fog of ballistic resistance. But ballistic resistance is worthless if the marine or soldier doesn't wear it."[3]

To put the 2007 time frame of designing the new helmet in context, military resistance to changing or upgrading pads was ongoing despite numerous complaints by troops to Doc Bob about wearability issues; pad testing was a concern of Congress; Doc Bob was fighting both the Marine Corps and the Army equipment bureaucracies over upgrading pads; the Blast Injury Research Program was just a start-up; and serious research on the extent of brain damage from exposure to blast had not begun. It was a time when the military equipment programs were not taking upgraded padded protection seriously or were just paying it lip service.

Despite emerging evidence of serious brain damage from exposure to blast, the Marine Corps moved forward in 2008 to develop the new helmet, called the Enhanced Combat Helmet (ECH), with only improved ballistic protection. Instead of a Kevlar shell, the new ECH would be made from a newly engineered thermoplastic material, called ultra-high molecular weight polyethylene (UHMWP). It meant that Kevlar, the primary material used to form the helmet shell since 1983, would no longer be used.

The Marine Corps partnered with the Army in developing the helmet, claiming that the helmet shell "exhibit[ed] an average of 35 percent improved ballistic protection as compared to the current state-of-the-art Army Combat Helmet."[4] Instead of upgrading the padded suspension system, the Marine Corps pointed out that the ECH would contain the existing padding that many marines were complaining about.

In December 2009 Doc Bob sought answers from the Marine Corps about why it had decided to retain the same padded suspension system for the new ECH, a system that marines constantly complained about because it caused discomfort and severe headaches while degrading performance. In response the ECH program manager at the time, Lt. Col. A. J. Pasajian (USMC) called Doc Bob to discuss the issue.

"In March 2009, we initiated the ECH program, and we got an urgent need that's driving it," Pasajian explained to Doc Bob. In typical military bureaucratic speak, Pasajian said, "Our focus is on Afghanistan as you see in the news today. I want you to know that the urgency of the moment is really a compelling issue here. So, for those reasons, we decided to keep the focus of this Enhanced Combat Helmet, to the helmet itself, to the raw materials we are investigating."[5]

Addressing the padded suspension system for the ECH, Pasajian told Doc Bob, "Also included is a completely new analysis on suspension and retention systems so we decided to keep those things separately. Now, we are doing that and it's called SMART-TE [Suspension Material Analysis and Retention Technology Test and Evaluation]." In trying to explain this "new analysis" but failing to make sense, he said,

> That effort, SMART-TE, is geared to two things: Number 1, we think we need to revise the test protocols to evaluate all suspension systems. So, whether it's pads or sling suspension system, a pad—sling hybrid and there's even something that's coming out of SBIRS [Small Business Innovation Research] that are unique in of themselves. They are not pads or slings. There are different things all together. There are hybrids and some things completely unique that we haven't seen before. There are some forms I can tell you that we have learned from our NATO partners that are unique in of themselves that are pretty interesting and I think present some considerable consideration.

Doc Bob, trying to steer Pasajian to a direct answer to why the Marine Corps was using the same pad system, said to Pasajian, "As you know, you look on my web site, I get hammered weekly by troops in combat in Afghanistan saying these damn things [stock-issued pads] detract from mission performance, and my thought was they got a rock in their shoe right now with those pads. If they buy a new shoe, the ECH, and put the same damn rock in it, it's going to hurt just as bad and is still not going to have optimum performance that the troops can feel comfortable doing what they have to do out in the field."

Although admitting their pads were not as comfortable as those issued by Operation Helmet, Pasajian said, "So, the performance is really driving our decision to stay with the current pad system and that is what we told the GAO and everyone else that's been asking us. I've been talking to a lot of people about this over the last couple of years and I got you, I understand you get a big, heavy vote."

Doc Bob mentioned the Lawrence Livermore research on the effects of the blast wave on the brain, the micro indentation of the skull itself, and the blast-wave effect being enhanced with a pad that is too firm. Pasajian would only say that they should look into it and include it in their analysis going forward. In other words the Marine Corps hadn't given the results any consideration.

Continuing the phone conversation, Doc Bob told Pasajian that the National Institute for the Blind seemed to be a roadblock in bringing new pads into the system. The NIB didn't seem to be very interested in looking at new materials, and it had the ability to say no, it was not going to work with certain vendors. He wanted to know if the Marine Corps had a role in the selection process. Pasajian replied that the Corps gave the NIB the specifications, and the NIB handled the contracting—a tacit acknowledgment that the NIB selected the vendors. Again, the military was denying involvement in the selection process. To Doc Bob his explanation about the pads did not make any sense. He said, "All they're doing is developing a new, updated helmet, but with the same rock in it."

However, while Pasajian was hyping to Doc Bob the urgency of producing the new helmet with upgraded ballistic protection, the development and fielding of the new combat helmet met a roadblock. Prototypes submitted by vendors in late 2009 failed to meet all the requirements, forcing the manufacturers to try again. Apparently, the prototypes "fell short" on ballistic and blunt-force tests, creating a setback to the ECH program.

In 2011 another setback occurred. At a House Armed Services Committee hearing in March 2011, commander of Marine Corps Systems Command Brig. Gen. Frank Kelley testified that the ECH was expected to fail its latest round of testing due to a production "anomaly," causing another delay in production. The "anomaly" was blamed on the manufacturer's production process, new project manager Lt. Col. Kevin Reilly told the *Army Times*. "The company altered the helmet's curing process," Reilly claimed. "After being cleared for low-rate production, the company had upped the temperature during the curing process to dry the paint faster." Reilly considered the problem as a "momentary setback for a helmet that will save lives."[6]

Just a few months later Marine Corps officials revealed to the *Army Times* that blaming the manufacturer for the test failures due to a production "anomaly" was incorrect. Instead, they admitted that the DOD testing procedures were to blame. Marine Corps testing determined that the curing was not the problem but blamed newly instituted DOD helmet-testing procedures, which were introduced just prior to the start of the helmet's First Article Testing. "Testing issues were numerous," Marine Corps officials told the *Army Times*. In fact problems revealed in testing procedures had a significant effect that could cause "a good helmet [to] fail" according to the officials.[7]

The officials expected that the helmet would not be ready for fielding until the first quarter of fiscal year 2014 (October 1, 2013–September 30, 2014), a good five years after the first projection of the "urgent need" requested in 2009. During late 2013 the Marine Corps announced it had overcome testing anomalies, and the helmet began to be fielded during the first quarter of 2014 with the same supply-issued pads used in the LWH.

In 2014 the Marine Corps published an ECH training video requiring that the pads be replaced after six months of continuous use. This requirement contrasts with the experience of marines who have asked Doc Bob for replacements of their more expensive Oregon Aero pads after three years of use in combat. Oregon Aero officials have seen them in use for as long as five years. It would seem, therefore, that using less expensive pads that have to be replaced after six months of use would, in the long run, be more expensive than initially purchasing pads that were costlier but lasted longer.

By 2016 the Marine Corps had fielded seventy-seven thousand ECH helmets and desired to field an additional 84,376. However, according to Military.com's gear blog of September 2016, the Marine Corps did not have the funding to purchase helmets and was relying on future funding through the fiscal year 2017 National Defense Authorization Act.[8]

Despite the development and fielding of the new ECH and the huge cost and effort involved, it was obsolete long before it went into production. The military bureaucracy did not take blast impact seriously, but by 2014 advancements in research disclosing serious degenerative brain damage from blast impact changed the thinking—internal cushioning from a padded suspension system may reduce impact damage to the brain. Pentagon officials finally acknowledged that the combat helmet being used by the troops was inadequate and that a helmet needed to be developed that also provided better protection from blunt-force and blast impact.

Instead, the Marine Corps announced in 2017 that it was purchasing a newer, lighter version of the ECH that "offered additional protection against small arms and frags" but made no mention of upgrading the padded suspension system to offer better protection against blast impact.[9]

It became a consensus among scientific helmet researchers that although the current combat helmet with a padded suspension system provided significant protection to the side and the back of the head, it did not protect the face and the soft tissue passages that allow the blast wave to reach the brain. This issue had been established years earlier by Canadian researchers, including Aris Makris, who design head protection for deminers.

In 2010 scientists at MIT's Institute for Soldier Nanotechnologies corroborated the Canadian research with their study of the effect of the ACH with a face shield on the propagation of blast waves within the brain tissue. They found that the face is the weakest link, causing the worst brain injuries. Like the Canadian researchers, they discovered that a face shield, along with the helmet, could significantly increase protection of soldiers from blast waves. From their testing they learned that the ACH without a face shield did not significantly mitigate blast effects into the intracranial cavity because it did not protect the face. Finding that pressure from blast can easily reach the brain through the soft tissue passageways of the eyes and the sinuses, they concluded that a face shield, as part of the combat helmet, would be significant in blocking that path.[10]

Also in 2010 Revision Military, formerly MSA (which bought out CGF Gallet), the Vermont-based manufacturer of protective military equipment, announced that it had designed a new lightweight helmet protection system with mandible guard and visor. The company called it Batlskin Head Protection System and claimed that it could help reduce brain injuries. "The Batlskin Head Protection System not only makes radical leaps forward in helmet and liner technology," CEO Jonathan Blanshay said. "Its integrated visor and mandible guard could also greatly reduce the incidence of traumatic brain injury (TBI) in blast situations." The use of foam padding was mentioned, but the company only revealed that it "use[d] dual foam technology for superior fit, comfort and impact absorption. Its multi-level design allow[ed] for cooling and stability while shim pads afford[ed] a custom fit."[11]

In 2013 the Army's Natick Research and Engineering Center began testing a storm-trooper-type helmet, similar to those used in Star Wars movies, that included improved ballistic and nonballistic impact protection. However, in 2014 the *Military Times* reported that the new helmet design had failed tests to decrease shock-wave pressure to the head. According to the report, the tests disclosed that blast waves could bounce off the added components and produce unexpected pressure. In addition, adding face protection didn't necessarily mean that it would lessen blast-wave impact.[12]

Then in January 2017, the Army began production of a base ECH that would have add-ons such as a visor, a jaw protector, and a "ballistic applique that serves as a protective layer." It resembles a motorcycle helmet, and the Army named it the Integrated Head Protection System. It is due to be fielded in 2020.[13]

However, the Naval Research Laboratory (NRL) introduced an unintended consequence when adding components to a helmet. Dr. David Mott, an aerospace engineer at NRL, said: "When you start adding these extra pieces of equipment, you don't always get what you expect. Multiple shocks interacting with each other can amplify the pressure, as can reflections off the structures that are in the suspension."[14]

Mott's experiments showed that a blast wave hitting the helmet with a padded suspension system reflects to the front of the helmet causing a "pressure spike on the forehead." When a face shield is added it keeps the blast wave from directly hitting the face, but a weaker wave does find its way to the face through the bottom of the shield. Adding a mandible shield with the face shield deflects most of the blast wave, but some of the wave still gets through a small gap below the mandible shield and reaches the face. However, part of the wave that hits the torso reflects to under the mandible shield combining with the other wave into one wave that is stronger than what hits the forehead if the face is not protected.

Mott explained that if the blast wave hits a helmet without a visor or mandible shield from behind, it escapes out the front of the helmet. Yet with a visor and mandible shield, the wave is trapped between the face and the visor and mandible shields, causing it to bounce around the head. Overall, Mott's research found that helmet blast response is complex. Mott noted, "An increase in the peak pressure in one location on the head is typically accompanied by reductions in other locations." The NRL continues to run experiments to better understand the physics of the blast wave effect on the helmet in order to design more protective helmets. In retrospect getting such a helmet to pass blast-pressure tests and deployed into the field would take a considerable amount of time, given how slow and cumbersome the bureaucracy works.

In order to develop and field a new helmet, the PEO Soldier Helmet Program Manager would contact Natick and determine what it has done in the area of a helmet with a face guard. Natick engineers would then design and develop prototypes and do human factors tests. Requirements and technical specifications would then be developed after which Natick would go out to industry and procure a limited quantity of prototypes.

In 2012 the Marine Corps's equipment bureaucracy finally began to recognize comfort as an important factor. The product manager for infantry combat equipment issued a request for information (RFI) soliciting information from manufacturers to procure an updated helmet padded suspension system called an Improved Helmet Suspension System (IHSS). According to the RFI, the product manager was interested in learning about new helmet suspension system technology that would improve blast and blunt-impact protection. Finally, the RFI stated that the IHSS "must provide stability and comfort for the helmet."[15]

In conjunction with the RFI, the Corps enlisted the help of the Applied Physics Laboratory (APL) at Johns Hopkins University to test and evaluate the new pad systems of manufacturers as part of the Marine Corps's SMART-TE program. The manager of the APL's Biomechanics and Injury Mitigation Systems Program reported that a helmet system "must also remain stable on the head during operations and maximize comfort to ensure proper use at all times."[16]

On the APL's website, Scott Swetz, the APL's project manager for SMART-TE, echoed Doc Bob's argument: "When we looked at existing helmet safety studies, we found that the majority considered only certain elements of protection and some bypassed important measures such as fit and comfort. If the helmet is not comfortable, it may not be worn properly and therefore won't provide the appropriate protection. The same goes for stability. If a helmet moves around during action, the warfighter could have compromised peripheral vision, be subjected to shrapnel and other dangers. All of these areas are equally important."[17]

However, wearability continued to be a problem with the ECH. During the first quarter of 2014 after the ECH was first fielded, Doc Bob reported that feedback from troops who used the new ECH prior to their deployment to Afghanistan revealed that the helmets were "horrible to wear." As one Army sergeant wrote: "We wear our ECH every day for 12 plus hours at a time. They are causing my guys to have headaches and they are constantly adjusting [their] helmets. This is detracting a significant amount of attention from security. Security is what keeps us alive. We cannot afford to have even a few seconds of distraction from this. We would really appreciate it if you would send us some better equipment so that my men could stay more focused on the mission and less occupied by unsatisfactory equipment."[18]

In July 2017, three years after the fielding of the ECH, Doc Bob still received emails from soldiers who were experiencing "massive headaches and discomfort" with the ECH using supply-issued pads, "taking away from situational awareness." It's the "old story of trying to fix rocks in your socks by changing the shoes," Doc Bob said.

Yet the military equipment bureaucracy continued to ignore the wearability issue. When Doc Bob asked troops why their complaints about the helmets wearability went unanswered, the near-unanimous response to them from senior NCOS was "stop being a bitch and suck it up." When Doc Bob asked senior NCOS the same question, their response was "funding."

Despite the wearability problem with the ECH, it was being lauded for its improved ballistic protection, and program managers were receiving awards from the Marine Corps for their management of the ECH program. According to the Marine Corps, the ECH was a "marked improvement" over the LWH in ballistic protection. However, no mention was made of any improvement in wearability or blunt-force or blast-impact protection.

As of 2016 upgrading helmet design to better protect against blast impact was still taking a backseat to another priority: a lighter helmet. In early 2016 a 10 percent lighter version of the ACH was fielded to soldiers. A redesign of the ACH to make it lighter also gave the Army the opportunity to upgrade blast protection and address pad comfort, but it did not do so, thus kicking that ball further down the road. Part of the hype publicizing the new, lighter version of the ACH was that it was more comfortable to wear. To a degree that is true. A lighter helmet is easier to wear, but the troops' biggest complaint about the helmet, at least to Doc Bob, was not the weight but the uncomfortable padded suspension system.

Then in early 2017, the Army, still preoccupied with a lighter helmet, announced a new generation of the ACH, called the ACH Gen II, that would reduce the weight an additional 24 percent. Again there was no mention of upgrading the padded suspension system. This new version of the ACH is not scheduled to be ready until 2022.[19]

With these new helmets, the problem of comprehensive protection from blast is still a problem. It's more than face protection that has some problems with the reflecting blast wave. It still comes down to the ability of the padded suspension system to absorb impact energy.

Doc Bob still believes he is doing the right thing no matter how long it takes: fighting with the military to provide the troops with the best padding available for both protection and comfort. It's why he has devoted more than thirteen years to this task. Yet the bureaucracy has continued to deny that there are any wearability problems with the pads. It has moved, at a snail's pace, from denying the advantage of pads—as the Marine Corps did—to redesigning helmets that better protect the troops from ballistics, an important issue, but still without any changes in the pads.

CHAPTER 29

The Bureaucratic Wall

We have met the enemy and he is us.

—Pogo

To change anything in the Navy is like punching a feather bed. You punch it with your right and you punch it with your left until you are finally exhausted, and then you find the damn bed just as it was before you started punching.

—Franklin D. Roosevelt, 1940

Roosevelt was assistant secretary of the Navy the first time he made these observations, and he was clearly frustrated with efforts to move the Navy bureaucracy to change. Decades later nothing has changed either in the Navy or in the military bureaucracy in general. In 2011 former secretary of defense Robert Gates on *60 Minutes* revealed his frustration with the military bureaucracy, stating that "a combination of stringent performance goals, inflexible rules and undetected 'culture creep' all conspire to leave behind a rigid organizational structure that can't switch directions when crises erupt. As a result, leaders have to go around the very bureaucracy that is supposed to help them, but turns out to get in their way."[1]

If a former president and a secretary of defense could not move the military bureaucracy, then what chance did outsiders like Jeff Kenner, Tammy Elshaug, and Doc Bob have in trying to effect change to help the troops? The Army's resistance to admit a problem with shorting the weave density of the Kevlar for the PASGT helmet led to justice being denied to Jeff and Tammy and perpetuated an ongoing risk to many soldiers. Both the Army's and the Marine Corps's stubborn resistance to changing their decisions and adopting a more wearable and protective helmet padded suspension system led to more than eighty-six thousand troops (as of 2018) requesting a better pad system from Operation Helmet rather than wear the supply-issued pads, which were both uncomfortable and caused many to suffer serious headaches that degraded performance in combat. Even an increase in incidents of TBI did not move the bureaucracy. Congress had to force the Army to enact a blast-injury research program to seek blast-injury mechanism solutions that could lead to an improved helmet design to mitigate the effects of TBI.

While Army and Marine Corps operating forces were changing the way they fight wars in the twenty-first century, with more flexibility, more creativity, and a quick reaction force, their support bureaucracies in the Pentagon still operate in a bubble, mired in a plodding twentieth-century business-as-usual inflexibility and resistance to change. Their failure to adapt to changing war needs resulted in a failure to meet the needs of the operating forces.

The military helmet bureaucracy's intransigence was just one example of a behavior pattern that also affected other combat protection equipment. An important example of that behavior was the change from the up-armor Humvee to the Marine Corps's Mine-Resistant Ambush-Protected (MRAP) vehicle to better protect troops from IEDs. This creative technology, requested by operating forces, met with bureaucratic intransigence that caused crucial delays, which put many lives at risk.

The primary problem with military equipment bureaucracies lies in their institutional thinking rooted in past projects. Once a bureaucracy takes ownership of a project and its requirements, it becomes entrenched and impervious to change. When confronted with changes from outside, the bureaucracy usually suppresses or rejects outright these types of changes to protect a current project and its funding and the survivability and careers of those responsible for it. The MRAP vehicle provides a prime example.

Franz Gayl is a science and technical expert who held the position of science and technology advisor for the Marine Corps Systems Command. With masters of science degrees in space systems opera-

tions and national resource strategy, he directed energy and nonlethal weapons for the Marine Corps and holds a patent for a nonlethal
weapon for immobilizing people. He is also a retired Marine who
spent twenty-two years on active duty, including a tour in Iraq. In
2010 Gayl, as a whistleblower, was stripped of his security clearances
and suspended indefinitely with pay after he revealed publicly in
2007 that the Marine Corps had quashed an urgent request from
senior officers in Iraq for the heavily armored MRAP vehicles to protect marines from the devastating effects of IEDs. Along with pub-
lishing a revealing case study on the MRAP issue, Gayl published two
case studies identifying two additional weapon systems, the Tactical
Concealed Video System and the Combat High Power Laser Daz-
zler, both important systems that would save lives but also suffered
from bureaucratic delays.[2]

For decades the successor to the jeep, the High Mobility Multi-
purpose Wheeled Vehicle, well known as the Humvee, or in military
language, the HMMWV, was the go-to vehicle for the Marine Corps.
Mainly a light transport vehicle with canvas doors, it could not withstand IED blasts and protect the marines riding inside it, but it was
an established program within the Marine Corps bureaucracy with
steady upgrades and constant funding. To Gayl the Humvee was "a
death trap." He revealed that for many years the Humvee bureaucracy
did not listen to advice given them about the vulnerabilities of the
vehicle. In fact they were aware of the MRAP vehicle and the threat
posed by IEDs years before the Iraq conflict but never took steps to
develop funding requirements to acquire the vehicle. According to a
2008 DOD inspector general audit report, the Marine Corps, despite
knowledge of the IED threat and the benefit of the MRAP vehicle to
counter it, "entered into operations in Iraq without having taken
available steps to acquire technology to mitigate the known mine
and IED risk to soldiers and Marines."[3]

Instead the Marine Corps Humvee bureaucracy decided to purchase
a factory-hardened Humvee, called the M1114, funded with $400 mil-
lion. Gayl noted that the MRAP vehicle was a large, heavily armored,
clunky vehicle that rode high off the ground, was almost laughable
in appearance, but was effective in protecting troops riding inside

it. He believed that the Marine Corps needed such a vehicle in Iraq, but the commanding general of the Marine Corps Combat Development Command rejected it. It was during the same time period that Marine Corps helmet officials rejected installing a padded suspension system for their new LWH. Gayl felt the Corps had rejected the MRAP because this "silly vehicle" presented a threat to the Humvee program and billions of dollars in funding.

According to Gayl in February 2005 the commanding general of the Marine Expeditionary Force in Anbar Province in Iraq was under fire because marines were getting blown apart every single day. Convinced that his troops could not continue to sustain such casualties, the general made an urgent request for mine-resistant vehicles, specifically the MRAP, to replace the Humvee and even provided the names of some of the manufacturers who produced the vehicle.

The implications of that one request, Gayl believed, was probably the biggest urgent need ever requested from the field in the Marine Corps. The request included 1111 vehicles with a price tag of about one and a half billion dollars. It was huge for the Marine Corps and a shock to the Humvee bureaucracy. The bureaucracy was obviously wrong about the upgraded Humvee, according to Gayl. Its concern was how to justify the Humvee and discourage the MRAP vehicle.

Much like the testimony in the 2006 hearing about helmet pads, the commandant of the Marine Corps testified before Congress to the durability of the Humvee, because of what the bureaucracy had told him. The request was a "not invented here" issue that put the bureaucracy, as Gayl put it, in a "real jam." It threatened theses bureaucrats' funding and their careers, so, Gayl said, they decided to convince the general in Iraq who had made the request that what he was really asking for was not an MRAP vehicle but just better protection from an upgraded Humvee.

Under Marine Corps protocol, the field commander was not supposed to ask for a specific solution. What the general actually wanted was the better protection that the MRAP vehicle provided. The M1114 had better protection than the older Humvee. To the bureaucracy the commander had essentially asked for better protection, and that was the story they would stick to. They resisted fulfilling the actual request because it would upset the bureaucracy funding. Instead of the MRAP, the commandant of the Marine Corps decided to replace all Humvees in theater with the M1114 up-armored Humvee. The deputy commandant of the Marine Corps for installations and logistics appealed to Marine Corps generals that the M1114 up-armored Humvee was the best available, most survivable asset to protect Marine Corps forces. So, Gayl felt, the whole story became contrived to avoid the MRAP vehicle in favor of the upgraded Humvee.

Thus the bureaucratic response to this major request was to risk everything to bury the true combat need, and according to Gayl "that is what they did." In August 2005 the Marine Corps Combat Development Command buried the MRAP vehicle request and did not develop a course of action for the urgent field request, attempt to obtain funding for it, or present it to the Marine Corps Requirements Oversight Council for a decision on acquiring an MRAP-type vehicle. This action resulted in a nineteen-month delay for an MRAP vehicle delivery during the height of the insurgency in Iraq. That delay, Gayl believed, resulted in hundreds of deaths and thousands of injuries.

It was done to protect the Humvee program, Gayl said, and the bureaucracy convinced Congress that it was the gold standard for protection. The MRAP vehicle request, if implemented, would also have disrupted other program bureaucracies such as those for the Expeditionary Fighting vehicle and the Light Armor Vehicle. Other programs would have had to come up with resources to pay for the request, and none of them was going to give up its money for a "not invented here" product.

From a bureaucratic standpoint, it was virtually impossible to buy armored vehicles off the shelf and send them to a combat zone. Even with congressional pressure, there would be massive resistance to change. Requirements needed to be established and operational testing conducted, but the Marine Corps was reluctant to face more operational testing that would reveal flaws in its weapons. These problems are often applied across the services. This reluctance to test was particularly strong for helmets and other combat gear that relate to the safety of the troops. In the end the bureaucracy was unwilling to adapt to the urgent request from the field commander to get MRAP vehicles in the field quickly.

It wasn't until July 2009, four years after the initial request for the MRAP vehicles, that more than thirteen thousand MRAP vehicles were fielded to marines and soldiers in Iraq and Afghanistan. It took pressure from the secretary of defense, Congress, warfighters, and journalists to overcome bureaucratic intransigence and get the MRAP vehicles produced and fielded. An investigation by the DOD inspector general resulted in a scathing indictment of the Marines Corps's failure to fulfill its fighters' 2005 need for the MRAP vehicles.

The MRAP vehicle fiasco is but one example of how entrenched military equipment bureaucracies do not willingly change themselves. It normally takes something drastic to force a change. The need for a more protective helmet using a padded suspension system should have been approved at the top levels of the Pentagon. According to Gayl,

> The assistant Marine Corps commandant . . . usually presides over this [kind of issue], approves it, and signs a paper that initiates the whole process with milestones. Now they have the ability to get money for purchasing a helmet. These gears are powerful gears that are set in motion. It takes on a legal momentum all on its own. It's extremely difficult to change, extremely exceptional, if it ever has changed that you had leaders with the guts and ability to stand up and say this is wrong path we are taking, we got to change. You can count those things throughout the history of the DOD on two hands. So, in other words, culturally no [change]. Complete inflexibility.

Once money is allocated for a helmet, including its suspension system, Gayl continued, "it's like a big rock rolling down the hill and for it to deflect left or right, it takes a lot of force if you can do it at all. For any shortcuts to the process to happen, it would violate the integrity of the program and the Marines are all about integrity—we do not do shortcuts because in the end it would hurt the marines in the field."

Gayl had considerable experience working with both military and civilian personnel within the Marine Corps Systems Command and the various bureaucracies that make up the command, including the equipment personnel. From his experience he opined that several factors could have contributed to the bureaucracy rejecting the purchase of the helmet pads that the marines in combat were requesting. One was that nobody from outside the bureaucracy was going to tell it which pads to buy; the bureaucracy would decide for itself. That is the crippling philosophy within the Systems Command.

Second, the bureaucracy doesn't care what pad it buys except that it's going to be cheap. It wasn't going to spend a lot of money on the pads, and the bureaucrats, according to Gayl, "hope it fails because it will prove [their] point in the first place that [they] didn't need pads. This is how the Marines operate." Gayl believes that Operation Helmet became the embarrassing enemy and extremely threatening to the stability of the bureaucracy and to the individuals whose reputations, careers, and staff promotions were on the line. People running the helmet bureaucracies needed to show that they weren't wrong and that they hadn't lied to the commander. That required them to weave a series of lies, obfuscations, and fantasies to prove Doc Bob wrong.

According to Gayl,

A whole bunch of interest grew up around the decision chain to approve the LWH helmet and make it a reality. There is the helmet project officer in which the helmet became his identity, his ownership. This is where the power of the Marines core personality comes from. They take ownership of it. They fall in love with it. There are books written about program managers falling in love with their programs and unable to change. That happens at all levels. I'm the project officer of the helmet and I'm hearing contrary views and maybe I'm a young guy, a captain or major, and I'm a little bit more flexible, or I'm a young civilian working for a senior civilian. There may be a little problem with the helmet. I'm hearing this and that. Then you get the arrogance of the higher level because he has the same problem in that he has taken ownership of the helmet. He's been advertising the helmet around this particular path. They haven't seen the product yet. It's on a set path, a basic design concept and it is firmly entrenched and it has been sold to the top leadership in the Pentagon. We want to purchase a helmet and field it to this requirement.

Also, Gayl explained the complexity of change—"to deviate from those programs requires a reprogramming of money and that's a big deal that somehow always attracts congressional attention. The whole process of change takes time that no one likes to do because it is hard going through all the bureaucratic layers."

The issue was really not whether the helmet was perfect when it was fielded. Rather Gayl felt that the helmet bureaucracy has "incentives that are much more similar to industry and that's why they are dangerous because they are incentivized by such things as fearing the loss of money, fearing the loss of status, jobs, etc." A good example is Natick's George Schultheiss, who insisted on procuring cheaper pads for the regular troops. Gayl noted, "It's hard to fire civilians; however, civilians can be made miserable."

Despite the Marine Corps equipment bureaucracy's initial and tenacious resistance to equipping helmets with the padded suspension system promoted by Doc Bob, it eventually decided to incorporate pads into its helmets. Gayl felt that decision might have been the result of the "forcing function" of being embarrassed publicly. "That's key," he said. "You have to embarrass them [the Marine Corps] from the outside. Humiliate them. Question their actual dedication from the outside. It's a horrible event for general[-level] officers to undergo in public. It must have been for General Catto. He probably went down and reamed some butts and then ended up forcing some issue, within the Marine Corps, to buy pads."

The patterns of behavior are very similar across the military services. The intensity in the Marine Corps reflects directly the intensity of the culture. "It's the strength of the culture," Gayl emphasized.

There's schizophrenia in the Marine Corps, Gayl explained.

Schizophrenia between the operating forces that we Americans associate with the Marine Corps and the support staff, which is mostly located in the Beltway—the Pentagon and Quantico, mostly populated with retired marines. That's where the problem begins and ends. The schizophrenic pieces, the commanding general that we always look to, to pin blame on, praise, are the holders of the legacy. When they are in their operating element, they understand everything about the organization. They were the individuals at one time who rose from the ranks. They understand the weapons, breakdown the weapons, and put them back together themselves. They know it all. The real masters of their art. Nothing gets by them. They understand everything at the technical and tactical levels, the leadership levels; they don't miss a thing. That's why the Marines are so effective and why the leadership works in the field. That's fine for the operating forces. They are the ones who take the initiative.

Once removed from the operating forces and placed in the bureaucracy, these warriors within a month are transformed schizophrenically into defenders of the existing program. "It takes no time at all," Gayle explained. "This is the Marine culture. You bloom where you are planted. No one in uniform really wants to get assigned to a support establishment. It's a grotesque job being taken out of the operating forces to work back at Quantico. But they go there, and they don't want to do a bad job. They will take ownership of whatever is handed to them, and they will make it work."

Gayl's seven-year battle with the Marine Corps to reinstate his security clearances and his job finally succeeded in September 2014. First, the U.S. Office of Special Counsel stymied the Marine Corps's attempts to end Gayl's career by ruling that he was a public whistleblower. Then in November 2011, a Navy review board, the Department of the Navy Central Adjudication Facility, overturned the Marine Corps's decision to strip Gayl of his security clearances, which led to cancellation of his suspension. Finally in September 2014, the Marine Corps and Gayl settled his whistleblower case through the Office of Special Counsel's mediation program. The decision prompted Gayl to say that he would return to the Marine Corps to work in support of all marines. According to USA Today, the Pentagon's Joint Program Office for MRAP vehicles has estimated that as many as forty thousand troops' lives were saved in Iraq and Afghanistan because of the MRAP vehicle.

Unfortunately, there is no place in a military bureaucracy for a forward, independent thinker like Gayl. For such a person, the bureaucracy will do everything possible to stymie, discredit, or remove him or her. Bureaucratic personnel are process oriented, expected to operate within predefined organizational rules, policies, and requirements, in a system where performance is often measured, not through initiative, creativity, or quality of work, but through meeting a set of arbitrary metrics. (The Veterans Administration bureaucracy is a good example of how these metrics do not work.) Personnel are not allowed to make subjective decisions—to make decisions that conflict with the rules and requirements.

Accountability for decisions such as those involving procurement, as Doc Bob, Oregon Aero, and others outside the system learned, are systemic in nature, not identifiable with any one individual or even an organization. Bureaucracies do not feel responsible for the needs or concerns of outsiders. Reality is defined only within the bureaucratic rules and requirements, and new realities and discoveries are not accepted within the thinking process. Officials operate within a bureaucratic bubble and often take the low-risk despite its effect on the troops. This is why it was so frustrating for Doc Bob, Oregon Aero, Jeff Kenner, Tammy Elshaug, and David Peterson to interact with the military bureaucracy, to make headway in changing the helmet suspension system or prosecute a fraud case.

Epilogue

We buy weapons that have less to do with battlefield realities than with our unending faith that advanced technology will ensure victory, and with the economic interests and political influence of contractors. This leaves us with expensive and delicate high-tech white elephants, while unglamorous but essential tools, from infantry rifles to armored personnel carriers, too often fail our troops

—James Fallows, "The Tragedy of the American Military"

As revealed in the preceding chapters, the military equipment bureaucracy has historically made acquisition decisions that have less to do with the battlefield and more to do with their obsession with preserving their individual programs and certain vendors over what is operationally needed in the field. Their efforts to maintain programs led them to resist change and dismiss outside innovations that could have helped protect troops in combat. These problems were pervasive throughout equipment and weapons innovation, design, and procurement, be it for a helmet or a fighter aircraft like the problem-plagued F-35.

The Marine Corps went out of its way to defend the disproved idea that the padless suspension system in its helmets was more protective than a padded suspension system. The Corps, in turn, resisted and attempted to discredit the work of Doc Bob and Operation Helmet to improve the protection and wearability of the helmets. The DOD, especially the Army, went so far as to rig blunt-impact tests for padded suspension systems in order to shut out a vendor that had what it called the "Cadillac of pads" in order to favor a less expen-

266

sive vendor because an equipment official believed that regular soldiers were not worth the extra costs of a better-quality pad system. This move caused thousands of soldiers and marines to complain to Doc Bob about the wearability of supply-issued pads, which often caused severe headaches.

These bureaucratic fiascos took place at the same time that TBI was named the signature injury of the war, mainly due to soldiers' and marines' exposure to IED blasts. The helmet issue was only part of a long history of the military bureaucracy's failure to provide protection for the troops that goes back to the American Civil War.

Doc Bob's valiant efforts cannot be underestimated, given his contribution to military awareness of the combat helmet's vulnerabilities in protecting its soldiers and marines from the devastating effects of TBI from exposure to blast. After failing to persuade the Marine Corps to upgrade its helmets to include a padded suspension system, he went public and testified before a congressional subcommittee about the benefits of using the padded suspension system. Before Doc Bob's activism and Operation Helmet began, the military had not formally recognized blast exposure as a cause of TBI or the important protective use of foam pads in combat helmets; the Marine Corps refused to use a padded suspension system in its helmets; and many of the troops still used the old PASGT helmet and its sling suspension system. Even now with both the Marine Corps and the Army using supply-issued pads, troops continue to request more comfortable pads. As recently as June 2018, Doc Bob received a request from a marine for 136 pads from Operation Helmet for marines that do patrols. Again, the important reason for the request was to reduce headaches so they can focus on the task at hand. Until the military takes wearability as seriously as protection, requests will continue to pour in to Doc Bob for pads that are more comfortable and more stable. Operation Helmet has now sent out over 92,600 Oregon Aero kits, so one would think these bureaucracies would suspect that Doc Bob was on to something important.

Doc Bob has been frustrated that he can't get the military bureaucracy to seriously discuss the realities of helmet wear versus laboratory testing. Many soldiers and marines told Doc Bob that they were unable to wear the helmets for the full duration of their mission because the supply-issued pads caused headaches so severe they could not concentrate on what they are doing.

Reflecting on more than thirteen years with Operation Helmet, Doc Bob explained that the military equipment bureaucracy was so intransigent because military members assigned to the Army's PEO Soldier and the Marine Corps Systems Command tend to be short-timers. By the time they find the restroom, they are headed for the next assignment. No matter how dedicated the military personnel may be to change, it's the civilian employees who wind up making the final determinations about what is wanted, needed, and procured.[1]

Also, the National Institute for the Blind had an inordinate role in deciding what equipment it would supply to the military, making decisions based on who would give it the most money for "co-production," even if it was only for stuffing foam in a bag or just shipping it to the various military departments. Unfortunately, Doc Bob learned that congressional oversight is woefully inadequate, "depending on whose goose is being gandered or which lobbyist is most favored."

"The issued pads cause headaches and a poor fitting helmet," a marine sniper wrote Doc Bob. "When the helmet doesn't fit right I cannot get the right check weld on my gun, decreasing accuracy and increasing risk.... The Ops-core helmets ... don't provide either the protection or fit.... The ECH, while a fine helmet to deflect small arms fire, still has the old, hard/uncomfortable Team Wendy pads. Why?"

"Why" is a constant question with Doc Bob and many of the troops. Helmets today are tested against the standards of earlier decades to prevent skull fracture from blunt-force impact.[2] But combat troops needed a helmet to protect them from blast and TBI. The equipment bureaucracy has failed to provide them with such a helmet or even an interim solution. The result is that more than 370,000 troops have been diagnosed with TBI, most often from exposure to blast.[3] The disconnect between the bureaucracy and the troops in combat has also caused Operation Helmet, as of mid-2018, to send over 92,600 helmet-pad upgrade kits to combat troops. This disconnect has plagued the troops through all the wars that the United States has been involved in. It is a disconnect brought on by a bureaucracy so set in its ways it will do anything to maintain its way of conducting business, including discrediting any outsider who dares to question its decisions.

But Pentagon officials, after more than fourteen years of war, are only now beginning to realize the high risk of severe brain damage that their service members face in these wars. The threshold for mild TBI is 50 g's, but according to the Army's Blast Injury Research group, blast-wave overpressure can accelerate the head up to 300 g's. Such impact on the brain makes it obvious that the primary protection for the men and women on the front lines—the helmet—is inadequate. The result should be a race to solve one of the military's biggest equipment challenges: developing a helmet that not only protects brains from ballistics and shrapnel but also from blunt-force and invisible, devastating blast-wave impact. But the first step in achieving this goal is to figure out the mechanisms linking blast exposure to TBI.

After more than ten years since its inception, the Blast Injury Research Group still does not fully understand the mechanisms leading to brain injury. Congress does not seem to be patient. In the 2018 National Defense Authorization Act, it mandated a "longitudinal [long term] medical study on blast pressure exposure of members of the Armed Forces during combat and training, including members who train with any high overpressure weapon system such as anti-tank recoilless rifles or heavy-caliber sniper rifles."[4] Also, in May 2018 U.S. Senator Joni Ernst (R-IA) proposed a bipartisan bill, "The Blast Exposure and Brain Injury Prevention Act of 2018," that requires the secretary of defense, among other things, to submit a plan improving research on and development of therapies for TBI.[5] It's difficult to know if these mandates will be any more effective than the last mandate issued in 2006. Given the DOD's slow and seemingly reluctant effort to come to a final conclusion on the matter, private research will need to lead the way, but the military bureaucracy, following its usual pattern, will most likely be reluctant to accept private research unless it's sanctioned by the military.

One entity is starting to recognize that a new helmet design is necessary to prevent soft-tissue brain damage such as CTE. In July 2017 Brown University reported it had received a $4.75 million grant from the Office of Naval Research to "develop new insights into how traumatic injuries form in the brain and develop new helmet technologies to help prevent them." Its research will "assess how well a helmet protects the soft tissue inside the skull—the brain and ultimately develop a prototype helmet that meets our new standard," including how much force inside the brain "is too much for cells."[6]

While the military has continued to use the same helmet pads since 2004, even in newly designed helmet shells, some pad manufactures are trying to come up with a new version of pad protection. One example is the vendor Angel 7 Industries, which developed a pad design that it provided to Doc Bob for his opinion. In December 2013 the company claimed on its website that independent testing of its helmet pads revealed "the extraordinary blunt-impact performance of the company's ASH-22 BioRmr padding system for combat helmets."[7] Doc Bob said the Angel 7 marketing director provided him with a sample of the company's pads, which he found interesting. The pads had a "dynamic suspension technology that is a single pad system that looks like an origami creature." Doc Bob gave the company's pads his "butt test," sitting on them in his office while he and the marketing director talked, something he has done with other pads. After about thirty minutes, he remarked to the Angel 7 representative, "These are too goddamn uncomfortable for my butt; they sure aren't going to work on your head."

Doc Bob believes in practical testing. Along with his "butt test," he has put pads in his ACH and worn the helmet for ten hours at a time around his house to determine their wearability. So far only Oregon Aero pads were comfortable enough to wear without distraction. The others were too uncomfortable and caused too much distraction. Another vendor, favored by the Marine Corps, developed an air-inflatable pad. But according to Doc Bob, "How in the hell are you going to carry a bicycle pump around with you to pump up your pads when you change altitude or temperature?"

The Army Research Laboratory in 2014 reported that it was developing new helmet-padding material that can withstand blast-pressure impact and protect the head from its acceleration. In doing so the ARL used computer modeling of head and helmet impacts to better understand how to tailor padding material. It has claimed that its research "is giving insight on optimal material structures and material combinations that achieve increased energy absorption while still being comfortable to wear."[8]

In addition in late 2017, ARL, teaming up with MIT, developed a tough synthetic-rubber material that can offer more ballistic strength in future combat helmets. The material is a type of highly resistant rubber made of polyurethane urea (PUU) that deforms to half its thickness when impacted at high speed and bounces back after impact. The current ECH used by the Army is made of high-performance UHMWPE plastic. The ARL's Dr. Alex Hsieh explained to ARL's Public Affairs that the new material showed "low resistance to elastic deformation under loading at ambient conditions and higher failure strain—the capability to sustain significantly greater amount of strain before failure—than most of the plastic materials."[9] This new material, primarily for ballistic protection, appears to resist backface deformation that can cause impact with the skull, leading to brain damage. However, it is not known how, or if, it would resist blast impact.

Brain injury and biomechanics expert John Lloyd, formerly a researcher with the U.S. Department of Veteran Affairs, learned through his research that "the manner in which the [foam pad] cells are arranged is particularly important" in mitigating rotational acceleration of blast impact. On September 3, 2015, his patent application was published for "Impact Absorbing Composite Material" that features a unique cell arrangement to reduce both linear and angular acceleration upon blunt-force and blast impact.[10]

The military's failure to properly protect the troops from TBI has resulted in enormous costs to taxpayers. According to the Congressional Budget Office, treating patients with brain injuries such as TBI and PTSD cost the Veterans Health Administration an estimated $2.2 billion during fiscal years 2004 through 2009. The International Hyperbaric Medical Foundation reported in 2011 that in ten years the total cost to taxpayers for such treatments is expected to be $1.35 trillion.

Oregon Aero is still supplying pads to the Special Forces for their MICH. Special Forces make their purchases under their own procurement group and not under an Army contract. The Army is still using

Team Wendy and MSA pads through the NIB under the Ability One program for its combat helmets.

However, Oregon Aero continues to be very active in sales of its other products to the military, primarily for flight helmets and the HALO and HAHO free-fall school. Crash-worthy pilot seats, especially Oregon Aero–designed ejection seats, for almost all aircraft in the Air Force inventory constitute much of its sales. Collaborating with the Air Force Research Laboratory, Oregon Aero also designed and manufactured ear seals as a helmet liner for the Joint Helmet-Mounted Cueing System used in most fast-moving jets. Oregon Aero continues to sell pad kits to Operation Helmet for troops wearing the ACH, the Marine Corps's LWH, and to those still using the PASGT helmet, and the pad kits are available to the public through the company's website.

Oregon Aero CEO Mike Dennis recalled that during the early years of the Marine Corps's fielding of the LWH to its combatants in Iraq, "the Marine Corps was processing ten traumatic brain injuries cases a day using their [unpadded] Lightweight Helmets. By their own estimate the cost was $3 million over the life of the injured soldier—$30 million dollars a day! This went on for two years before the Marines stopped making the information available to the public. Sadly these are just the numbers; the personal disaster far exceeds this. It was shameful."[11]

Mike believes the helmet bureaucracy is impenetrable. He said he learned that the "best performing gizmo is not going to make the grade with the military unless you get control of the money and the political process. It's not the technical expertise that counts. It's the political process that's vital despite the quality of the gizmo." Oregon Aero is now primarily a research and development company. CEO Mike Dennis said, "We're still here, spinning our understanding of non-Newtonian physics to solve other pressing troubles." The company no longer has "time to grind on a grudge."

During the time that the Sioux Manufacturing case was being litigated in 2007 for shorting the weave of Kevlar panels for the PASGT helmet, another company was in the process of producing defective ACHS and LWHS that contained adulterated and weakened Kevlar sold to the DOD. Two whistleblowers, who were supervising prison-inmate workers at the U.S. Federal Penitentiary at Beaumont, Texas (operating under the name UNICOR), revealed that defense contractor and helmet manufacturer ArmorSource (formerly Rabintex USA) while overseeing the manufacture of ACHS and LWHS at its subcontractor UNICOR intentionally removed strands of Kevlar from the woven panels in order to install them more quickly into the helmets. Ohio-based ArmorSource is a private company established in 2005. In 2006 ArmorSource became an official supplier of the ACH to the DOD.

Removing Kevlar strands reduced the weave density below the critical minimum, thus weakening protection for the troops wearing the helmets. Also, Rabintex USA shipped rejected and adulterated helmets to the DOD, which also putt troops at risk wearing them. They took old Kevlar panels that were in poor, brittle condition and pounded them with heavy objects to make them pliable enough to fit into the helmets being manufactured. At that time Sioux Manufacturing Corporation was producing Kevlar panels for ACHS as a subcontractor to UNICOR.

After the whistleblowers notified the DOD of the problem, the Army responded (unlike in the Sioux Manufacturing case) in May 2010 by recalling 44,000 ACHS already fielded, another 55,000 helmets in storage, and did not accept 3,000 others from the company for a total of 102,000 helmets affected. The DOJ was subsequently notified of the case, and on September 22, 2010, it filed a *qui tam* civil fraud complaint against ArmorSource and its parent company, Rabintex Industries Ltd., in the federal district court in Beaumont, Texas.

Ironically, the Army conducted ballistics tests that resulted in the helmets failing the tests. In the Sioux Manufacturing case, the PASGT helmet supposedly passed ballistics tests based on the fact that projectiles did not completely penetrate the helmet, but the Army ignored the backface deformation effect caused by a projectile not fully penetrating the helmet but penetrating enough to cause the inner lining to bulge and strike the wearer's skull. In the Rabintex case, the failure of ballistics tests was this time based on the backface response of the helmets. The Army's chief of testing reported that the ballistic tests demonstrated that the helmets would leave the solider unprotected against a "worst-case scenario" strike on the helmet.

An investigation by the DOJ-OIG along with the Defense Criminal Investigative Service determined that UNICOR's manufacturing problems led to defective ACHS and LWHS that were not manufactured in accordance with contract specifications. They found that UNICOR was also preselecting helmets for inspection and testing and had falsely certified that the helmets passed inspections and met specifications.[12]

A surprise inspection by DOJ-OIG found that the contractor, Armor-Source, did not provide adequate oversight of the work. Also, the investigators uncovered evidence that inmates were openly using improvised tools on the ACHS, damaging the ballistic material and creating the potential for tools to be used as weapons in the prison. The Defense Contract Management Agency, the DOD's watchdog at a contractor site, failed to uncover the problems. The DOJ settled the case, with ArmorSource agreeing to pay $3 million. Jeff's and Tammy's exposé helped prompt a closer look at these helmet manufacturers, and although their case was compromised, they set in motion investigations that exposed a bigger problem and led to a vital recall.

More than four years after signing the settlement agreement, Jeff Kenner tries not to let the case bother him despite all that he has endured. However, it still affects him when he runs into former coworkers and says hello to them and they don't say hello back. Devils Lake is a small community, and it's common to encounter people who work at Sioux Manufacturing. Some are generally as nice as they always have been, but others ignore him. If they ever become aware of the full account of what happened, maybe they will come around. Jeff can now just hand them this book.

Jeff is disappointed in the Army. He thought it was on his side and that it would want to give the best possible equipment to the soldiers. It seems to him that even now the Army is "crapping on the soldiers when they come back from Iraq and Afghanistan." He is frustrated that "Congress gets way more benefits, are set for life, while soldiers are coming back injured and not taken care of."[13] It bothers Jeff that some of the helmets may have failed and soldiers were injured or killed as a result. He feels bad about the lawsuit, especially because it had to be directed against the whole tribe, as the owner of Sioux Manufacturing. Also, despite doing good work at the company, Jeff was then treated badly, and that has been tough to swallow. His is a classic whistleblower case, where the company wants to kill the messenger rather than solve the problem. Jeff now drives a refrigerated semi-truck for Levers Foods, delivering mostly milk and ice cream to grocery-store chains in North Dakota and Minnesota.

Tammy Elshaug believes that Sioux Manufacturing got off pretty easy, but she has said she would do it all over again under the same circumstances. Jeff feels guilty that he didn't come forward sooner, but he didn't really understand the significance of the problem at first. He thinks the DOJ did a really good job but believes the Army really downplayed the problem to avoid embarrassment and having to admit that any soldiers were hurt or even killed because of its negligence. Additionally, the Army ignored the results of the DOJ investigation and justified its decision to continue to allow UNICOR to award contracts to Sioux Manufacturing, stating that the company "has not admitted or acknowledged any wrongdoing with respect to its contract performance."[14]

Tammy also runs into former coworkers who ignore her. A few, however, are supportive and think she did the right thing. To this day she has nightmares about Sioux Manufacturing, that she is still working there or has to go back to the plant for some reason and the company won't let her do her job. She then wakes up and wonders how she could ever dream such a thing. Perhaps it is because she spent over half her life there. Although she hasn't heard anything about the alleged affair anymore, she wonders what people think. "They look at you funny and you know what they are thinking, but you brush it off and keep going," she lamented.[15]

Tammy is currently employed with the U.S. Post Office in Devils Lake, sorting mail as a clerk. She is making more money at the post office after two years than she made at Sioux Manufacturing after twenty-six years.

David Peterson's only reflection on the case is that from the initial filing, it concerned him greatly and personally because the Kevlar pattern sets were going into the helmets that young men and women were wearing in Iraq, Afghanistan, and wherever troops were deployed. He felt that if the allegations were correct and the underweaving was a significant safety issue, then the case was important and deserved the fullest investigation, which he and his colleagues tried to give it.

Ultimately, Peterson thinks that, given the information provided and the fact that the helmets, at least according to the military's "experts," passed all the ballistics tests, the case ultimately became a contract issue, which he and his colleagues tried to settle as favorably as possible. Other factors such as the issuing of a new contract with Sioux Manufacturing also had a significant impact on the settlement amount. Peterson thinks that Jeff and Tammy performed an appropriate service for the troops but were not treated well by Sioux Manufacturing, which he believes spread false rumors of an illicit relationship between them.

Peterson did not become involved in another *qui tam* case after the Sioux Manufacturing matter. On December 31, 2008, at the age of sixty-seven, he retired from the U.S. attorney's office. After retirement he took on the position of state president of AARP in North Dakota.

Sioux Manufacturing has shifted its production to the manufacturing of ablator tiles, a type of heat shield, used for missiles. Former CEO Carl McKay died of cancer in February 2018.

The Army and the Marine Corps have moved on in helmet-shell design from Kevlar to plastic, in use now with the ECH, that affords the wearer more ballistic protection. However, at the same time there has been no advancement in developing foam pads, and no requirement stipulates that helmet design must protect against blast impact. This situation is no doubt due to the Blast Injury Research Program's failure to reach a consensus agreement on the mechanisms of blast-induced TBI. Thus a requirement for blast mitigation is still a long way off.

In May 2018 researchers at the Center for a New American Security published a study, primarily for the Army science and technology community, acknowledging that military helmets only partially mitigate the effects of blast impact and calling for establishing "an interim requirement for protection against blast overpressure while continuing further research to refine the requirement over time."[16]

With all that, the cost to taxpayers is small compared to the suffering of the troops, who assumed that they would be sent into battle with the best equipment that the United States can offer—especially the helmet that was protecting the most vulnerable part of the body, the brain. Despite the unyielding military bureaucracy, our country can surely do better. Hopefully, more people will pick up the challenge that Doc Bob, Tammy, and Jeff started and finally get the troops a helmet that is worthy of their bravery and sacrifice.

Acknowledgments

We would not have been able to write this book without the help of many individuals who gave their time and shared information. We especially want to thank Jeff Kenner, Tammy Elshaug, Dr. Robert Meaders, Mike Dennis, and Pierre Sprey, who gave of their time and the valuable information that brought the book to life. A special thanks to Art Levine, who did much of the original investigations on the helmet pads and was amazingly generous with his knowledge and his research. We also would like to thank Cher (yes that Cher) for supporting Robert Meaders and his Operation Helmet to help save troops from brain injuries.

Many thanks to our literary agent, Philip Turner, who came to us to write this book. He stayed with us throughout this journey and has been very helpful with his wisdom, engaging the publisher, the University of Nebraska Press, and moving us forward when necessary to completion.

We want to thank the University of Nebraska Press for taking on this book and being so helpful in its publication. We would especially like to thank Natalie O'Neal, editorial assistant; Joeth Zucco, senior project editor; Barbara Wojhoski, freelance copyeditor; Ann F. Baker, editorial, design, and production manager; and Tish Fobben, direct response manager, Marketing Department.

Individual thanks are also in order, first from Robert:

I am thankful for the many professional colleagues and associates who were helpful during my career as a criminal investigator for the Department of Defense. I am grateful to the many friends and business associates who made my post-government career a success. I am especially thankful, given my "senior" age, to have maintained the good health and wellness that have allowed me to pursue endurance running and many marathon and half-marathon races.

I am thankful for the support of my family during the long, difficult, and time-consuming process of researching and writing this book: to my wife, Norma, who endured many absences while I traveled throughout the country and spent extensive time on the book; to my two sons, Blake and Marc, and their wives, Debbie and Debbie, for their wonderful support; and finally, to my lovely grandchildren, Sophia, Luke, Emily, Daniel, Nick, and Eleanor, for the pleasures they bring to my life every day.

And now thanks from Dina:

I would like to thank all the professional friends over the years who taught me how to investigate and help whistleblowers. My gratitude goes to my first mentor, Ernest Fitzgerald, who at age ninety-two can still occasionally show me his wicked sense of humor. I would also like to thank my Project on Government Oversight (POGO) family. I started POGO in 1981 and now serve as the treasurer of the board. I especially want to thank the executive director, Danielle Brian, who started at POGO as an intern and became one of my dearest friends—she is the kind of person who, no matter how busy she is, will find the time to encourage me and comfort me through all our stages of life. I also want to thank Keith Rutter, whom I hired decades ago. Keith helped guide a little guerilla organization into an oversight powerhouse in Washington. He unfortunately also had to organize me, which was probably his greatest and most difficult task. He has made POGO into family, and Danielle and Keith have kept the organization young and on the cutting edge—something hard to do in a nonprofit that is decades old.

I would especially like to thank Pierre Sprey, who was willing to teach me, an inexperienced twenty-five-year-old investigator, how to read weapons test reports and beat the bureaucracies at their own game. It is hard to believe that in this book, we are still at it thirty-seven years later!

I have had the help of many reporters in different projects over the years, but Jonathan Alter as always been there in the past years to give great advice and compare notes on life and survival.

I want to thank Dr. Jonathan Terdiman of the University of California, San Francisco and Dr. Karimi Gituma of Forward Medicine for patching me together over time so that I can work and be useful.

The San Francisco City Chorus, under the direction of Larry Marietta and accompanist John Walko, has given me a wonderful break every week from the sometimes-dark world of investigation to make beautiful classical music and clear my soul. Special thanks to Larry M for being the most fun and the best director I have ever had.

I have a great group of friends scattered all over the country who have been very important in my professional and private life journey: Donna Martin, who was my first investigator at POGO and is a wonderful lifelong friend; Leslie Ferguson and Helen Read, who are great friends and also my next-door neighbors; Frank Thomas, who became a good friend and taught me how to sing correctly for the first time in my life; Janet Michaelis, who was my sister's best friend, but we realized that we are both cut from the same muckraker cloth, and she has become a good friend to plot with; my college friend, Bob Eicholz, and his husband, Steve Scott, who give excellent advice about life—so glad that Bob and I reconnected after many years—Tim McCune, who was my employee and now another friend whom I can call on for good and balanced advice; Greg Williams, another former employee who is now a friend who has followed his bliss; and Matt Renner, an extremely talented man who has been so helpful to me in my current career and took the time to be a good friend. (Keep an eye on young Matt; I am sure he will emerge as one of the new national leaders.)

I am blessed to still have both of my parents, Ned and Genny Rasor, who at ninety-two and married for seventy-one years continue to inspire me. I caught their passion for politics, and they sealed the deal of me wanting to be an investigator and government reformer when they took me to the Watergate hearings when I was in high school. I am also lucky to have my talented scientist/sculptor sister, Julia Rasor, who listens to me wail when I think life is unfair.

I have two wonderful sons, Daniel and Nicholas Lawson, who fill me with pride as they excel in their careers and are so fun to be with. I greatly admire Nick's girlfriend, Robin Chung, for her many talents and her good heart toward the less fortunate.

Acknowledgments

My remarkable husband of thirty-eight years, Thom Lawson, doesn't want me to mention him because he likes to do wonderful things without getting the credit, so I will only say what I said in my last book: he has known me since I was sixteen and married me anyway. Much of my life's success has been because of his great and steady support when I thought I could not accomplish my goals. Any woman should be so lucky to find such a wonderful life partner.

Notes

Introduction

1. Justin Meaders, interview by Robert Bauman and Dina Rasor, December 9, 2009.

2. Justin Meaders, interview by Robert Bauman, June 22, 2010.

3. Justin Meaders, interview by Bauman and Rasor, December 9, 2009.

4. Justin Meaders, interview by Bauman and Rasor, December 9, 2009.

5. Terri Tanielian and Lisa Jaycox, eds., *Invisible Wounds of War: Psychological and Cognitive Injuries, Their Consequences, and Services to Assist Recovery* (Santa Monica CA: Rand Corporation Center for Military Health Policy Research, 2008).

6. Conn Hallinan, "The Traumatic Brain Injury Epidemic," *CounterPunch*, June 21, 2011, https://www.counterpunch.org/2011/06/21/the-traumatic-brain-injury-epidemic/.

7. Jennifer Chu, "Modeling Shockwaves through the Brain," *MIT News*, September 29, 2014, http://news.mit.edu/2014/modeling-shockwaves-through-brain-0929.

8. Jerry A. Boriskin, "Ambiguity Colliding with the Unknown: Complex PTSD + Addiction + Traumatic Brain Injury," Veterans Administration Northern California Health Care System, Mare Island, California, Powerpoint Presentation, date unknown.

9. Dr. Segun Toyin Dawodu, "Traumatic Brain Injury (TBI)—Definition and Pathophysiology," Medscape, September 22, 2015, https://emedicine.medscape.com/article/326510-overview.

10. Charles W. Hoge, Dennis McGurk, Jeffery L. Thomas, Anthony L. Cox, Charles C. Engel, and Carl A. Castro, "Mild Traumatic Brain Injury in U.S. Soldiers Returning from Iraq," *New England Journal of Medicine*, January 31, 2008, http://www.nejm.org.

11. Edgar Jones, Nicola T. Fear, and Simon Wessely, "Shell Shock and Mild Traumatic Brain Injury: A Historical Review," *American Journal of Psychiatry*, November 2007 164, no. 11: 1641–45, http://dx.doi.org/10.1176/appi.ajp.2007.07071180.

12. Charles W. Houff and Joseph P. Delaney, "Historical Documentation of the Infantry Helmet Research and Development," U.S. Army Material Command Five-Year Personnel Armor System Technical Plan, Human Engineering Laboratory, Aberdeen Proving Ground, Maryland, February 1973.

13. Christian Beekman, "Combat Helmets Have Moved beyond Just Protection," *Task & Purpose*, June 12, 2015, https://taskandpurpose.com/combat-helmets-have-moved-beyond-just-protection/.

14. Shailesh G. Ganpule, Linxia Gu, Aaron L. Alai, and Namas Chandra, "Role of Helmet in the Mechanics of Shock Wave Propagation under Blast Loading Conditions," *Mechanical and Materials Engineering Faculty Publications* 56 (January 1, 2011), http://digitalcommons.unl.edu/mechengfacpub/56.

15. Christine Anderson, "FDA Surveillance Threatened Whistleblowers," Project on Government Oversight (POGO) blog, February 28, 2014, http://www.pogo.org/blog/2014/02/fda -surveillance-threatened-whistleblowers.html.

16. Jillian Berman, "Company Retaliation against Whistleblowers Rises to All-Time High, Survey Finds," Huffington Post, Business, January 12, 2012, http://www.huffingtonpost .com/2012/01/06/business-ethics-_n_1189110.html.

1. Hard-Headed Marines

1. Justin Meaders, interview by Robert Bauman and Dina Rasor, December 9, 2009.

2. Justin Meaders, interview by Bauman and Rasor, December 9, 2009.

3. Justin Meaders, interview by Bauman and Rasor, December 9, 2009.

4. Justin Meaders, interview by Bauman and Rasor, December 9, 2009.

5. Justin Meaders, interview by Bauman and Rasor, December 9, 2009.

6. Dr. Robert Meaders, interview by Robert Bauman and Dina Rasor, December 8, 2009.

7. Dr. Robert Meaders, interview by Bauman and Rasor, December 8, 2009.

8. Aris Makris, Vice-President Research and Development, Chief Technical Officer, Allen-Vanguard, Med-Eng Systems, Ottawa, Canada, email to Dr. Robert Meaders, May 2, 2005.

9. "New Helmet Improves Protection," U.S. Army Soldier and Biological Chemical Command, U.S. Army Soldier Systems Center—Natick MA, Press Release, November 14, 2000, http://www.sbccom.army.mil.

10. Dr. Robert Meaders, interview by Bauman and Rasor, December 9, 2009.

11. Dr. Robert Meaders, email to U.S. Marine Corps Headquarters, subject: Head armor: Does the Commandant know the new helmets have the same defect as the PASGT for non-ballistic impact protection?, citing position of LTCOL Patricio, USMC, LWH Program Manager, March 30, 2005.

12. Dr. Robert Meaders, interview by Bauman and Rasor, December 9, 2009.

13. Dr. Robert Meaders, interview by Bauman and Rasor, December 9, 2009.

2. Twisted Logic

1. Christopher C. Trumble (T.R.U.E. Research), B. Joseph McEntire and S. Crowley (USAARL), "Blunt Head Injury Protection for Paratroopers. Part II: Improved System Description," USAARL Report no. 2005-05, Aircrew Protection Division, March 2005.

2. "Impact Attenuation, Marine Corps Helmets, MICH (TC-2000), Lightweight Helmet (LWH), PASGT," Marine Corps Team, Natick Soldier Center, Natick MA, August 1, 2003.

3. Dr. Robert Meaders, interview by Bauman and Rasor, December 9, 2009.

4. Dr. Robert Meaders, email to U.S. Marine Corps Headquarters, subject: Head Armor: Does the Commandant know the new helmets have the same defect as the PASGT for non-ballistic impact protection?, citing position of Lieutenant Colonel Patricio, USMC, LWH Program Manager, March 30, 2005.

5. Dr. Robert Meaders, interview by Bauman and Rasor, December 9, 2009.

6. Dr. Robert Meaders, email to USMC commandant reproducing helmet test results from Intertek Testing Services NA Inc., Cortland NY, March 30, 2005.

7. D. M. Fitzgerald, Program Manager, Infantry Combat Equipment, U.S. Marine Corps Systems Command, "Lightweight Helmet (LWH) Sling Suspension vs. Ballistic Liner and Suspension System (BLSS)," U.S. Marine Corps, Marine Corps Systems Command, Quantico VA, April 14, 2005.

8. Fitzgerald, "Lightweight Helmet (LWH) Sling Suspension."

9. Fitzgerald, "Lightweight Helmet (LWH) Sling Suspension."

10. Fitzgerald, "Lightweight Helmet (LWH) Sling Suspension vs. Ballistic Liner and Suspension System (BLSS)."

11. G. R. Cox, Acting Medical Officer of the Marine Corps Memorandum to Commanding General, Marine Corps Systems Command, Third endorsement on MARCORSYSCOM letter, Subject: "Lightweight Helmet (LWH) Sling Suspension vs. Ballistic Liner and Suspension System (BLSS)," April 25, 2005.

12. Cox, Third endorsement on MARCORSYSCOM letter.

13. Dr. Robert Meaders, interview by Bauman and Rasor, December 9, 2009.

14. Brian T. Hart, "Part I: Ringing Ears Hear No Evil: Marine SYSCOM's Lightweight Approach to Head Injury, *Minstrel Boy* (blog), July 2, 2006, https://minstrelboy.blogspot.com/2006/07/part-i-ringing-ears-hear-no-evil.html.

15. Dr. Robert Meaders, interview by Bauman and Rasor, December 9, 2009. Dr. Meaders produced numerous emails from marines in combat.

16. "New Helmet Improves Protection," U.S. Army Soldier and Biological Chemical Command, U.S. Army Soldier Systems Center—Natick MA, Press Release, November 14, 2000.

17. "New Helmet Improves Protection."

18. Letter from Dr. Robert Meaders to Maj. Gen. William Catto, USMC, November 14, 2005.

19. R. Meaders, Operation Helmet blog, 2004–8.

20. Dr. Robert Meaders, letter to Major General Catto, Marine Corps Systems Command, Quantico VA, November 14, 2005.

21. Meaders letter to Catto.

22. Meaders letter to Catto.

23. Meaders letter to Catto.

24. Justin Meaders, interview by Bauman and Rasor, December 9, 2009.

3. Moment of Truth

1. Unless otherwise noted, recollections and quotations in this chapter are from Jeff Kenner, interviews by Robert Bauman and Dina Rasor, December 2, 2008.

2. Recollections and quotations here and in the next four paragraphs are from Tammy Elshaug, interview by Robert Bauman and Dina Rasor, December 2, 2008.

3. Sioux Manufacturing human resource manager (name withheld at his request), interview by Robert Bauman, October 31, 2009

4. Blowback

1. Kenner, interview by Bauman and Rasor, December 2, 2008.

2. Jeff Kenner, email to Carl McKay, subject: "Thank You," September 22, 2004.

3. Carl McKay, email to Jeff Kenner, subject: "RE: Thank You," September 22, 2004.

4. Sioux Manufacturing human resource manager, interview by Bauman, October 31, 2009.

5. Kenner, interview by Bauman and Rasor, December 2, 2008.

6. Tammy Elshaug, interview by Bauman and Rasor, December 2, 2008.

7. Tammy Elshaug, interview by Bauman and Rasor, December 2, 2008.

8. Recollections and quotations here and in the next three paragraphs are from Kenner and Tammy Elshaug, interviews by Bauman and Rasor, December 2, 2008.

5. Whistleblower's Nightmare

1. Larry Elshaug, interview by Robert Bauman, September 9, 2009.

2. Tammy Elshaug, interview by Bauman and Rasor, December 2, 2008.

3. Tammy Elshaug, interview by Bauman and Rasor, December 2, 2008.

4. Kenner, interview by Bauman and Rasor, December 2, 2008.

5. Kenner, interview by Bauman and Rasor, December 2, 2008.

6. Recollections and quotations here and in the next six paragraphs are from Kenner and Tammy Elshaug interviews by Bauman and Rasor, December 2, 2008, and September 16, 2009.

7. Daniel Herman, fire chief, Fort Totten Rural Fire Department, letter to whom it may concern, October 31, 2005.

8. Commentary by Spirit Lake Tribe and Sioux Manufacturing Corporation sent to the *New York Times* concerning the newspaper's article of February 6, 2008.

9. Sioux Manufacturing human resource manager, interview by Bauman, October 31, 2009.

10. Kenner, interview by Bauman and Rasor, December 2, 2008.

11. Tammy Elshaug, interview by Bauman and Rasor, December 2, 2008.

12. Sioux Manufacturing human resource manager, interview by Bauman, October 31, 2009.

13. Sioux Manufacturing human resource manager, interview by Bauman, October 31, 2009.

14. Hyllis Dauphinais, interview by Robert Bauman, December 1, 2011.

15. Carl McKay, interview by Robert Bauman, October 26, 2011.

16. Kenner and Tammy Elshaug, interviews by Bauman and Rasor, December 2, 2008.

6. The Tinkerer and His Unique Foam

1. Recollections and quotations in this chapter are from Mike Dennis, interview by Robert Bauman and Dina Rasor, June 29, 2009. Also some of the information in this chapter was provided by a confidential source familiar with CGF Gallet.

7. Mr. Helmet

1. Pierre Sprey, interview by Art Levine and Mishi Ibriham for Dan Rather production, 2009.

2. Sprey, interview by Levine and Ibriham, 2009.

3. Confidential source, interview by Robert Bauman and Dina Rasor, September 2, 2009.

8. Shut Out

1. Unless otherwise indicated, recollections and quotations in this chapter are from Dennis, interview by Bauman and Rasor, June 29, 2009

2. Dr. Robert Meaders, interview by Bauman and Rasor, December 9, 2009, and confidential source, interview by Bauman and Rasor, September 2, 2009.

9. Cost More Important Than the Troops?

1. Unless otherwise noted, recollections and quotations in this chapter are from Tony Erickson, COO, Oregon Aero, interview by Art Levine, April 8, 2008, and John Rice, former salesman, Oregon Aero, interview by Art Levine, 2008.

2. Numerous attempts by the authors to provide Schultheiss an opportunity to comment on the information cited in this book regarding his actions resulted in his continued declination to be interviewed. He died in February 2017.

3. Sprey, interview by Levine and Ibriham, 2009.

4. Sprey, interview by Levine and Ibriham, 2009.

5. Sprey, interview by Levine and Ibriham, 2009.

6. Dr. Robert Meaders, interview by Bauman and Rasor, December 9, 2009.

10. Search for Justice

1. Recollections and quotations in this chapter are from Kenner and Tammy Elshaug, interviews by Bauman and Rasor, December 2, 2008; and Andrew Campanelli, interview by Robert Bauman and Dina Rasor, February 17, 2009.

11. The New York Lawyer

1. Recollections and quotations in this chapter are from Campanelli, interview by Bauman and Rasor, February 17, 2009.

12. Operation Helmet Becomes a Force

1. Franz Gayl, interview by Robert Bauman, July 23, 2011.

2. Unless otherwise indicated, recollections and quotations here and in the following paragraph are from Robert Meaders, interview by Bauman and Rasor, December 8, 2009.

3. "Marines Corps Helmets, MICH (TC-2000), Lightweight Helmet (LWH), PASGT," Marine Corps Team, Natick Soldier Center, Natick MA 01760, August 1, 2003.

4. Mark Meaders, interview by Robert Bauman and Dina Rasor, February 2009.

5. "Traumatic Brain Injury (TBI): Management of Mild to Moderate TBI in the Iraqi Theater of Operations: Lessons and Observations on Practices, Training, and Equipment," U.S. Marine Corps Center for Lessons Learned, April–May 2006, http://mcwl.marines.mil.

13. Feds Move In

1. David Peterson, interview by Robert Bauman, September 15, 2009.

2. Recollections and quotations here and in the following four paragraphs are from Peterson, interview by Bauman, September 15, 2009; Kenner and Tammy Elshaug, interviews by Bauman and Rasor, December 2, 2008; and Campanelli, interview by Bauman and Rasor, February 17, 2009.

3. Campanelli, interview by Bauman and Rasor, February 17, 2009.

4. Peterson, interview by Bauman, September 15, 2009.

5. Peterson, interview by Bauman, September 15, 2009.

6. Campanelli, interview by Bauman and Rasor, February 17, 2009.

7. Campanelli, interview by Bauman and Rasor, February 17, 2009.

8. Peterson, interview by Bauman, September 15, 2009.

9. Campanelli, interview by Bauman and Rasor, February 17, 2009.

10. Kenner and Tammy Elshaug interview by Bauman and Rasor, December 2, 2008.

11. Tammy Elshaug, interview by Bauman and Rasor, December 2, 2008.

12. Peterson, interview by Bauman, September 15, 2009.

13. Kenner, interview by Bauman and Rasor, December 2, 2008.

14. Peterson, interview by Bauman, September 15, 2009.

15. Kenner and Tammy Elshaug, interviews by Bauman and Rasor, December 2, 2008.

16. Peterson, interview by Bauman, September 15, 2009.

14. Breakthrough for the Troops

1. Dr. Robert Meaders, interview by Bauman and Rasor, December 8, 2009.

2. Unless otherwise indicated, the description of and quotations from the hearing are from "Transcript of House Armed Services Committee: Subcommittee on Tactical Air and Land Hearing on Combat Equipment in Iraq and Afghanistan," June 15, 2006.

3. Dr. Robert Meaders, interview by Bauman and Rasor, December 8, 2009.

4. Dr. Robert Meaders, interview by Bauman and Rasor, December 8, 2009.

5. Rep. Curt Eldon, Chairman Tactical Air and Forces Subcommittee letter to Honorable Kenneth J. Krieg, June 19, 2006.

15. Small Victory, Large Defeat

1. Justin Meaders, interview by Bauman and Rasor, December 9, 2009.

2. Marine Corps Headquarters Memorandum, "Pad Suspension System in the Lightweight Helmet," 2006.

3. Dr. Robert Meaders, interview by Bauman and Rasor, December 8, 2009.

4. Quotations here and in the next two paragraphs are from Justin Meaders, interview by Bauman and Rasor, December 9, 2009.

5. Recollections and quotations here and in the next five paragraphs from Dr. Robert Meaders, interview by Bauman and Rasor, December 8, 2009.

6. GAO Report GAO-09-768R, "Warfighter Support: Information on Army and Marine Corps Ground Combat Helmet Pads," July 28, 2009, https://www.gao.gov/products/GAO-09-768R.

16. Army Resistance

1. Unless otherwise noted, quotations and recollections in this chapter are from Peterson interview by Bauman, September 15, 2009.

2. Recollections and quotation here and in the following paragraph are from Campanelli, interview by Bauman and Rasor, February 17, 2009; and U.S. Department of Justice, Office of the Inspector General, Report of Investigation, Subject: Sioux Manufacturing Company, January 15, 2008.

3. Recollections and quotations here are from Campanelli, interview by Bauman and Rasor, February 17, 2009.

17. Cut and Paste

1. Kenneth J. Krieg, Undersecretary of Defense for Acquisition, Technology and Logistics, letter to the Honorable Solomon P. Ortiz, Chairman, Readiness Subcommittee, Committee on Armed Services, U.S. House of Representatives, Washington DC, and to the Chairman and Ranking Member of the Air and Land Forces Subcommittee, February 22, 2007.

2. Unless otherwise indicated, the following narrative, including quotations, is based on Art Levine, interview by Robert Bauman, May 3, 2012. Levine provided the authors with his in-depth notes, interviews, and other documents of his investigation into helmet pads.

3. Quotations here and in the following paragraph are from Philip E. Coyle testimony before the Committee on Armed Services, House of Representatives, 110th Congress, Department of Defense Body Armor Programs, June 6, 2007.

4. Roger Charles, "Army Acquisition Capos Fraudulently Alter Body Armor Test Results," *Defense Review*, June 23, 2008.

5. "Warfighter Support, Independent Expert Assessment of Army Body Armor Test Results and Procedures Needed before Fielding," GAO Report 10-119 to Congressional Requesters, October 2009.

6. James Fallows, "M-16: A Horror Story," *Atlantic*, July 1981, https://www.theatlantic.com/magazine/archive/1981/06/m-16-a-bureaucratic-horror-story/545153/. See also Fallows, *National Defense* (New York: Random House, 1981).

7. Pierre Sprey, interview by Robert Bauman and Dina Rasor, May 1, 2012.

8. Robert Scales, "Gun Trouble," *Atlantic*, January–February 2015, https://www.theatlantic.com/magazine/archive/2015/01/gun-trouble/383508/.

18. Damage Control

1. Unless otherwise noted, recollections and quotations in this chapter are from Robert Meaders, interview by Bauman and Rasor, December 8, 2009. Also, Levine, interview by Bauman, May 3, 2012, and Sprey, interview by Bauman and Rasor, May 1, 2012.

2. Dr. Robert Meaders, letter to the Honorable Kenneth J. Krieg, March 19, 2007.

19. Negotiation

1. Recollections and quotations in this chapter are from Peterson, interview by Bauman, September 15, 2009; Campanelli, interview by Bauman and Rasor, February 17 2009; and Kenner and Tammy Elshaug, interviews by Bauman and Rasor, December 2, 2008, and by Bauman, September 16, 2009.

20. Settlement

1. Recollections and quotations in this chapter are from Peterson interview by Bauman, September 15, 2009; Campanelli, interview by Bauman and Rasor, February 17, 2009; and Kenner and Tammy Elshaug, interviews by Bauman and Rasor, December 2, 2008, and by Bauman, September 16, 2009.

21. Just Give Me What I Want

1. George Schultheiss, email to Oregon Aero, subject: MFF Liner, February 2007.

2. Tony Erickson, COO, Oregon Aero, email to George Schultheiss, subject: RE: MFF Liner, February 22, 2007.

3. Lee Owen and Tony Erickson, Oregon Aero telephone meeting with George Schultheiss, February 2007.

4. Lee Owen and Tony Erickson, interviews by Art Levine, 2008.

5. Sprey, interview by Bauman and Rasor, May 1, 2012.

6. Erickson, interview by Levine, 2008.

7. Erickson, interview by Levine, 2008.

22. Kafkaesque Nightmare

1. Recollections and quotations in this chapter are from Brig. Gen. Mike Caldwell, commanding officer, Oregon National Guard, interview by Dina Rasor, February 12, 2010, and by Art Levine, 2008. Also Frank Frysiek, Col. Dave Ferre, U.S. National Guard, and Dave Greenwood, U.S. National Guard, and Lt. Col. Robert Preiss, legislative liaison, U.S. National Guard, interviews by Art Levine, September 2008.

23. Fallout

1. "Sioux Manufacturing Settles Dept. of Justice Case," *Devils Lake Journal*, December 19, 2007.

2. Joseph Marks, "Spirit Lake Leaders Dispute New York Times Story," *Grand Forks Herald*, February 14, 2008.

3. "Sioux Manufacturing Settles Dept. of Justice Case."

4. Campanelli, interview by Bauman and Rasor, February 17, 2009; "Sioux Manufacturing Corporation Pays $1,935,000 to Settle False Claims Allegations," Department of Justice, U.S. Attorney Drew H. Wrigley, District of North Dakota, press release, December 18, 2007.

5. Quotations here and in the following paragraph from Bruce Lambert, "Manufacturer in $2 Million Accord with U.S. on Deficient Kevlar in Military Helmets, *New York Times*, February 6, 2008, https://www.nytimes.com/2008/02/06/us/06helmet.html.

6. "Crew and VoteVets.org Demand Investigation into $74 Million DOD Contract," CREW, https://www.citizensforethics.org/press-release/crew-and-votevets-org-demand-investigation-into-74-million-dod-contract/, accessed July 4, 2018.

7. Senator John F. Kerry, letter to DOD Inspector General Claude M. Kicklighter, February 8, 2008.

8. Myra Pearson, Chair, Spirit Lake Nation, letter to Melanie Sloan, Executive Director, CREW, February 12, 2008.

9. Pearson, letter to Sloan, February 12, 2008.

10. Quotations here and in the next three paragraphs from "Commentary by Spirit Lake Tribe and Sioux Manufacturing Corporation concerning *New York Times* Article of February 6, 2008," enclosed with Pearson letter to CREW, February 12, 2008.

11. "Commentary by Spirit Lake Tribe."

12. Richard Squillochoti, Army Research Laboratory, interview by Robert H. Bauman, September 2009.

13. Trina Gooding, Defense Supply Center, Philadelphia, email to Robert H. Bauman, September 3, 2009.

14. Squillochoti, interview by Bauman, September 2009.

15. Campanelli, interview by Bauman and Rasor, February 17, 2009.

16. "Dorgan Says Pentagon Should Investigate ND Armor," kxnet.com North Dakota News Network, February 28, 2008.

17. Peterson, interview by Bauman, September 15, 2009.

18. "CREW and VoteVets.org Demand Investigation into $74 Million DOD Contract," CREW, https://www.citizensforethics.org/press-release/crew-and-votevets-org-demand-investigation-into-74-million-dod-contract/, accessed July 4, 2018.

24. Delamination

1. Sara Campbell, "PASGT Helmet Test: An Example of Effective Intra-Government Testing Collaboration," *International Test and Evaluation Association Journal* 30 (December 2009): 557–61.

2. J. van Hoof and M. J. Worswick, Department of Mechanical Engineering, University of Waterloo, Waterloo ON, "Combining Head Models with Composite Models to Simulate Ballistic Impacts," Defence R&D Canada, Defence Research Establishment Valcartier Contract Report, June 2001, p. 1.

3. Van Hoof and Worswick, "Combining Head Models," p. 1.

4. "Testing of Body Armor Materials: Phase III" (Washington DC: National Academies Press, 2012), chap. 7, "Helmet Testing," 142.

5. Narrative and quotations here and in the next four paragraphs from Sprey, interview by Bauman and Rasor, May 1, 2012.

6. Quotations here and in the following paragraph from Dixie Hisley, U.S. Army Research Laboratory, Army Evaluation Center, Aberdeen Proving Ground, Maryland, response to written questions from Robert Bauman, April 11, 2012.

25. Won't Work Unless You Wear It

1. Shared by Dr. Robert Meaders, interview by Bauman and Rasor, December 8, 2009.

2. Recollections and quotations here and in the next paragraph from Justin Meaders, interview by Bauman and Rasor, December 9, 2009.

3. Shared by Dr. Robert Meaders, email to Bauman and Rasor, April 3, 2010.

4. Dr. Robert Meaders, email to Bob Fellers, Dick Vetter, Jack Visage, Rod McMahon, subject: PEO Soldier Meeting, May 12, 2011.

5. Quotations here and in the next paragraph from Dennis, interview by Bauman and Rasor, June 29, 2009.

6. Shared by Dr. Robert Meaders, interview by Bauman and Rasor, December 8, 2009.

7. Operation Helmet Troop Feedback Report, March 26, 2007, shared with authors by Dr. Robert Meaders.

8. Performance Specification for Helmet, Advanced Combat, U.S. Army Research Development and Engineering Command, Natick Soldier Center, Natick MA, October 6, 2005.

9. Quotation here and in the next paragraph from Robert Meaders, telephone interview by Bauman, June 2017.

10. Marine Corps Systems Command, "Improved Helmet Suspension System Industry Day," Power Point slides, Product Manager, Infantry Combat Equipment, October 17, 2012.

11. Quotations here and in the next paragraph from Stephanie Desmon, "Soldiers Evacuated with Headaches Don't Return," JHU Gazette, November 14, 2011.

12. Comments from Operation Helmet website, https://operation-helmet.org.

13. Combat Helmet Pad Suspension Systems Limited Field User Evaluation and Operational Assessment, DOT&E Summary and Findings, February 2009.

14. Quotations here and in the next four paragraphs from Sprey, interview by Levine and Ibriham, 2009.

15. Dr. Robert Meaders, interview by Bauman and Rasor, December 8, 2009.

16. Brian J. Ivins, "How Satisfied Are Soldiers with Their Ballistic Helmets? A Comparison of Soldiers' Opinions about the Advanced Combat Helmet and the Personal Armor System for Ground Troops Helmet," Military Medicine 172, no. 6 (2007): 586–91, here 586.

17. Bob Meaders, "2011 Industry Day," Doc Bob's Blog, July 26, 2011, https://www.operation-helmet.org/2011-industry-day/.

18. "Uncomfortable Combat Helmet: Deal with It," Stand for the Troops (blog), June 11, 2011, http://sftt.org/wendy-combat-helmet/.

19. "Helmet Headache," Army Times, June 25, 2007.

20. Quotation here and in the following paragraph from John L. Sweeny, "Helmet Pad Debate," Army Times letters, July 7, 2007.

21. Dr. Robert Meaders, email to Robert Bauman, August 1, 2011.

22. Sprey, interview by Levine and Ibriham, 2009.

23. Quotations here and in the next six paragraphs are from Dan Rather Reports, "Operation Helmet," episode 416, May 2009.

24. Recollections and quotations here and in the next paragraph are from Dr. Robert Meaders, interviews with Robert Bauman, 2015–17.

26. Intransigent

1. Unless otherwise noted, recollections and quotations are from Dr. Robert Meaders, interview by Bauman and Rasor, December 8, 2009.

2. "Unauthorized Procurement of Ballistic Protection, Body Armor, and other Safety Items," Army Directive from Department of the Army, Washington DC, to All Army Activities, April 17, 2009.

27. Too Little, Too Late

1. Sig Christenson, "Grim Toll of Military Suicides Reaches a New Milestone," *San Antonio Express-News*, April 8, 2016.

2. Jeffery Nerenberg, Jean-Philippe Dionne, and Aris Makris, Med-Eng Systems, Inc., "PPE: Effective Protection for Deminers," *Journal of Mine Action* 7, no. 1 (2003): 47–51.

3. Aris Makris, J. Nerenberg, J. P. Dionne, Med-Systems Inc., C. R. Bass, Automotive Safety Laboratory, University of Virginia, C. Chichester, U.S. Army, "Reduction of Blast Induced Head Acceleration in the Field of Anti-Personnel Mine Clearance," abstract, date unknown, available at http://www.dtic.mil/dtic/tr/fulltext/u2/a458451.pdf.

4. Quotations here and in the next two paragraphs from Aris Makris, email to Dr. Robert Meaders, subject: RE: Head Protection (brain injury), May 2, 2005.

5. Michelle K. Nyein, Amanda M. Jason, Li Yu, Claudio M. Pita, John D. Joannopoulos, David F. Moore, and Raul A. Radovitzky, "In Silico Investigation of Intracranial Blast Mitigation with Relevance to Military Traumatic Brain Injury," Proceedings of the National Academy of Sciences of the United States of America, November 30, 2010, http://www.dtic.mil/dtic/tr/fulltext/u2/a533075.pdf.

6. Paul Scharre and Lauren Fish, "Reform Weapons Training to Protect US Troops from Brain Injury," The Hill, May 7, 2018, http://thehill.com/opinion/national-security/386468-reform-weapons-training-to-protect-us-troops-from-brain-injury.

7. Ronald Glasser, "A Shock Wave of Brain Injuries," *Washington Post*, April 8, 2007, http://www.washingtonpost.com/wp-dyn/content/article/2007/04/06/AR2007040601821.html?noredirect=on.

8. "Carson Study: 1in 6 Shows TBI Symptoms," *Army Times*, posted by the Associated Press, April 11, 2007, http://www.armytimes.com/news/2007/04/ap_carson_braininjury_070411/.

9. Tanielian and Jaycox, *Invisible Wounds of War*.

10. "Traumatic Brain Injury (TBI)."

11. John Lloyd "Military Helmets May Provide Little Protection against Traumatic Brain Injury," paper accepted for presentation at the Military Health System Research Symposium, Fort Lauderdale FL, August 2014.

12. Department of Defense Directive 6025.21E, "Medical Research for Prevention, Mitigation, and Treatment of Blast Injuries," July 5, 2006, http://www.esd.whs.mil/Portals/54/Documents/DD/issuances/dodd/602521p.pdf.

13. Blast Injury Research Program Background, http://blastinjuryresearch.amedd.army.mil.

14. "Prevention, Mitigation, and Treatment of Blast Injuries: FY12 Report to the Executive Agent," Department of Defense Blast Injury Research Program Coordinating Office, U.S. Army Medical Research and Materiel Command, http://blastinjuryresearch.amedd.army.mil.

15. "Prevention, Mitigation, and Treatment of Blast Injuries: FY14 Report to the Executive Agent," Department of Defense Blast Injury Research Program Coordinating Office, U.S. Army Medical Research and Materiel Command. http://blastinjuryresearch.amedd.army.mil.

16. Jon Hamilton, "Pentagon Shelves Blast Gauges Meant to Detect Battlefield Brain Injuries," NPR Now, December 20, 2016, https://www.npr.org/podcasts/500005/npr-news-now.

17. Hamilton, "Pentagon Shelves Blast Gauges."

18. Elana Duffy, "DoD Canceled Brain Injury Research Because It Didn't Like the Results," Task & Purpose, December 23, 2016, http://taskandpurpose.com/author/elana-duffy/.

19. Michael J. Leggieri, "Blast-Related Traumatic Brain Injury Research Gaps," IEEE Xplore Digital Library, June 19, 2009, http://ieeexplore.ieee.org.

20. Michael J. Leggieri, in response to questions submitted by Robert H. Bauman, January 29, 2010.

21. Michelle K. Nyein, Amanda M. Jason, Li Yu, Claudio M. Pita, John D. Joannopoulos, David F. Moore, and Raul A. Radovitzky, "In Silico Investigation of Intracranial Blast Mitigation with Relevance to Military Traumatic Brain Injury," Proceedings of the National Academy of Sciences of the United States of America, January 4, 2011, http://pnas.org.

22. Donna Miles, "Research Examines Blast Impact on Human Brain," Defense.gov News, April 12, 2011.

23. "Prevention, Mitigation, and Treatment of Blast Injuries: FY12 Report to the Executive Agent."

24. "Research Raises Concerns for New Army Helmet Design," Army Times, August 24, 2014.

25. Sprey, interview by Bauman and Rasor, May 1, 2012.

26. Recollections here and in the next seven paragraphs are from William Moss and Michael King, Lawrence Livermore National Laboratory, interview by Robert Bauman and Dina Rasor, February 2010.

27. William Moss, Michael King, and Eric Blackman, "Skull Flexure from Blast Waves: A Mechanism for Brain Injury with Implications for Helmet Design," Physical Review Letters 103, no. 108702 (September 4, 2009), http://www.pas.rochester.edu/~blackman/mkb09.pdf.

28. Moss, King, and Blackman, "Skull Flexure from Blast Waves."

29. Eric Blackman, interview by Robert Bauman, February 24, 2010.

30. Dina Rasor, "People First . . . and Dogs, Too: A Case Study of Throwing Money and High Technology at a Military Problem," Truthout, May 26, 2011, https://truthout.org/articles/people-first-and-dogs-too-a-case-study-of-throwing-money-and-high-technology-at-a-military-problem/.

31. Miles, "Research Examines Blast Impact."

32. "Prevention, Mitigation, and Treatment of Blast Injuries: FY15 Report to the Executive Agent," Department of Defense Blast Injury Research Program Coordinating Office, U.S. Army Medical Research and Materiel Command, http://blastinjuryresearch.amedd.army.mil.

33. William Moss and Michael King, Lawrence Livermore National Laboratory, interview by Robert Bauman and Dina Rasor, July 7, 2011.

34. Dr. Robert Meaders, email and telephone interview with Robert Bauman, July 2011.

35. Moss and King, interview by Bauman and Rasor, July 7, 2011.

36. James Dao, "Brain Ailments in Veterans Likened to Those in Athletes," New York Times, May 16, 2012, https://www.nytimes.com/2012/05/17/us/brain-disease-is-found-in-veterans-exposed-to-bombs.html.

37. Seth Robbins, "Doctors Study Link between Combat and Brain Disease," Stars and Stripes, January 23, 2010, https://www.stripes.com/news/doctors-study-link-between-combat-and-brain-disease-1.98394.

38. Dr. Bennet Omalu, Dr. Jennifer L. Hammers, and Dr. Julian Bailes, "Chronic Traumatic Encephalopathy in an Iraqi War Veteran with Posttraumatic Stress Disorder Who

Committed Suicide," *Neurosurgical Focus* 31, no. 5 (November 2011), http://thejns.org/doi/full/10.3171/2011.9.FOCUS11178.

39. Ann C. McKee and Meghan E. Robinson, "Military-Related Traumatic Brain Injury and Neurodegeneration," HHS Public Access, author manuscript, December 4, 2014, http://transcraniallighttherapy.com/wp-content/uploads/2016/10/Military-related-traumatic-brain-injury3.pdf.

40. Lee E. Goldstein, Andrew M. Fisher, Chad A. Tagge, Xiao-Lei Zhang, Libor Velisek, John A. Sullivan, Chirag Upreti, et al., "Chronic Traumatic Encephalopathy in Blast-Exposed Military Veterans and a Blast Neurtrauma Mouse Model," HHS Public Access, author manuscript, August 9, 2013, https://www.ncbi.nlm.nih.gov/pmc/articles/PMC3739428/.

41. Robert F. Worth, "What if PTSD Is More Physical Than Psychological?," *New York Times*, June 10, 2016, /https://www.nytimes.com/2016/06/12/magazine/what-if-ptsd-is-more-physical-than-psychological.html.

42. "Prevention, Mitigation, and Treatment of Blast Injuries: FY16 Report to the Executive Agent," Department of Defense Blast Injury Research Program Coordinating Office, U.S. Army Medical Research and Materiel Command, http://blastinjuryresearch.amedd.army.mil.

43. "Prevention, Mitigation, and Treatment of Blast Injuries: FY16 Report to the Executive Agent."

44. Ann C. McKee, Nigel J. Cairns, Dennis W. Dickson, Rebecca D. Folkerth, C. Dirk Keene, Irene Litvan, Daniel P. Perl, Thor D. Stein, Jean-Paul Vonsattel, William Stewart, et al., "The First NINDS/NIBIB Consensus Meeting to Define Neuropathological Criteria for the Diagnosis of Chronic Traumatic Encephalopathy," *Acta Neuropathologica* 131, no.1 (January 2016): 75–86, https://www.ncbi.nlm.nih.gov/pmc/articles/PMC4698281/.

45. Dr. Sharon Baughman Shively, Dr. Iren Horkayne-Szakaly, Dr. Robert V. Jones, Dr. James Kelly, Dr. Regina C. Armstrong, and Dr. Daniel P. Perl, "Characterization of Interface Astroglial Scarring in the Human Brain after Blast Exposure: A Post-mortem Case Series," *Lancet Neurology*, June 9, 2016, https://doi.org/10.1016/S1474-4422(16)30057-6.

46. Shively et al., "Characterization of Interface Astroglial Scarring."

47. Dr. Bennet I. Omalu, Dr. David X. Cifu, Dr. Leonard E. Egede, and Dr. Uchenna S. Uchendu, "National Expert Panel on TBI and Chronic Traumatic Encephalopathy Morbidity and Mortality among Vulnerable Veterans," Focus on Health Equity and Action, U.S. Department of Veterans Affairs, June 30, 2016, https://www.hsrd.research.va.gov/.

48. Anna Almendrala, "Scientists May Be on the Way to Developing a Test for CTE," *Huffington Post*, September 26, 2017, updated September 27, 2017, https://www.huffingtonpost.com/entry/cte-diagnose-test_us_59ca984be4b07e9ca11f3877.

49. "First-Ever Blood Test for Detecting Brain Injury Cleared by FDA," Health.mil, March 15, 2018, http://www.health.mil.

50. "Combat Veterans Coming Home with CTE," CBS News, *60 Minutes*, January 7, 2018.

51. Luke Ryan, "TBIS and PTSD: The conflation between the two," *SOFREP News*, January 18, 2018, https://sofrep.com/98240/depression-among-veterans/.

52. "How IEDs May Be Physically Causing PTSD," CBS News, *60 Minutes*, April 1, 2018.

53. "Prevention, Mitigation, and Treatment of Blast Injuries: FY16 Report to the Executive Agent."

28. Back to the Future

1. Eric G. Blackman, Melina E. Hale, and Sarah H. Lisanby, "Improving TBI Protection Measures and Standards for Combat Helmets," Semantic Scholar, University of Rochester, 2011, http://www.pas.rochester.edu/~blackman/tbihelmets07new.pdf.

2. Blackman, Hale, and Lisanby, "Improving TBI Protection Measures."

3. Dennis, interview by Bauman and Rasor, June 29, 2009.

4. Dan Lamothe, "MC and Army Plan to Roll Out New Battlefield Helmet," *Marine Corps Times*, April 20, 2009.

5. Recollections and quotations here and in the next five paragraphs are from Dr. Robert Meaders, interview by Bauman and Rasor, December 8, 2009.

6. Tony Lombardo, "Enhanced Combat Helmet Has Been in Development for Two Years," *Army Times*, March 27, 2011.

7. James K. Sanborn, "Twice-Delayed Next-Gen Helmet Gets New Tests," *Army Times*, November 17, 2011.

8. Hope Hodge Seck, "Corps Needs Cash to Field New Helmet to All Marines," Military.com, September 14, 2016, https://www.military.com/kitup/2016/09/corps-wants-field -enhanced-combat-helmet-marines.html.

9. James Clark, "The Marine Corps Goes Light with New Body Armor, Plates, Packs, and Helmets," *Task & Purpose*, July 17, 2017, https://taskandpurpose.com/marine-corps-new -body-armor-plates-packs-helmets/.

10. Gordon Gibb, "Would Face Shield Protect Soldiers from Brain Injury?," Lawyersand-Settlements.com, December 2, 2010, https://www.lawyersandsettlements.com/articles/brain _injury/brain-injury-traumatic-9-15509.html.

11. David Crane, "Revision Military BATLSKIN Modular Head Protection System (MHPS) Lightweight Ballistic Combat Helmet, Visor and Mandibular Guard/Ballistic Face Shield: Complete Ballistic Maxillofacial Protection (Facial Armor) for the 21st Century Warfighter," *Defense Review*, September 14, 2010, http://www.defensereview.com /revision-military-batlskin-modular-head-protection-system-mhps-ballistic-visor-and -mandibular-guardballistic-face-shield-complete-ballistic-maxillofacial-protection-for -the-21st-century-warfighter/.

12. Kris Osborn, "Report: New Army, Marine Helmet Fails Test," Military.com, August 29, 2014.

13. Sarah Sicard, "The Army Just Started Producing Its Brand New Ballistics Helmet," *Task & Purpose*, March 23, 2017, https://taskandpurpose.com/army-just-started-producing -brand-new-ballistics-helmet/.

14. Explanation and quotes here and in the next two paragraphs are from "NRL Simulates IED-Like Blast Waves Against Army Helmet Prototypes," U.S. Naval Research Laboratory, July 17, 2014, https://www.nrl.navy.mil/news/releases/nrl-simulates-ied-blast-waves -against-army-helmet-prototypes.

15. "RFI-Improved Helmet Suspension System," Solicitation Number M67854-12-1-1079, U.S. Marine Corps System Command, September 2012.

16. "Laboratory Setting Standards for 'SMART' Helmets," Johns Hopkins University Applied Physics Laboratory, May 9, 2012, http://www.jhuapl.edu/newscenter/stories/st120509.asp.

17. "Laboratory Setting Standards for 'SMART' Helmets."

18. Dr. Robert Meaders, email posted on Operation-Helmet.org website.

19. Alex Horton, "Lighter Combat Helmets in Store for Army," *Stars and Stripes*, March 22, 2017, https://www.stripes.com/lighter-combat-helmets-in-store-for-army-1.460033.

29. The Bureaucratic Wall

1. Jena McGregor, "Gates' *60 Minutes* interview: Why Should Leaders Have to Create Bureaucratic Workarounds in Their Own Organizations?," *Washington Post*, May 16, 2011.

2. Unless otherwise noted, recollections and quotations here and in the remainder of the chapter are from Franz Gayl, interview by Robert Bauman, July 23, 2011.

3. "Marine Corps Implementation of the Urgent Universal Needs Process for Mine Resistant Ambush Protected Vehicles," Inspector General, United States Department of Defense Report no. D-2009-030, December 8, 2008.

Epilogue

1. Unless otherwise noted, recollections and quotations in the chapter are from Dr. Robert Meaders, interviews with and emails to the authors.

2. "New Grant Supports Comprehensive Research on Traumatic Brain Injury," Brown University Public Release, July 26, 2017, https://news.brown.edu/articles/2017/07/tbi.

3. "Ernst Fights to Prevent Traumatic Brain Injuries among U.S. Soldiers," *Ripon Advance*, May 22, 2018, https://riponadvance.com/stories/ernst-fights-prevent-traumatic-brain-injuries -among-u-s-soldiers/.

4. "Longitudinal Medical Study on Blast Pressure Exposure of Members of the Armed Forces," National Defense Authorization Act, H.R. 2810 Sec. 734, 2017.

5. "Ernst Fights to Prevent Traumatic Brain Injuries."

6. "New Grant Supports Comprehensive Research."

7. "A7 Helmet Systems Validates Blunt-Impact Performance of Its ASH-22 BioRmr Suspension and Padding System," Angel 7 Industries, December 20, 2013, http://angel7industries .com/A7-2013-PR.pdf.

8. "3D Printed Synthetic Bones May Help Prevent Wartime Brain Injuries," Engineering.com 3D Printing, News and Commentary for Professionals, August 12, 2014, https:// www.engineering.com/3DPrinting/3DPrintingArticles/ArticleID/8236/3D-Printed-Synthetic -Bones-May-Help-Prevent-Wartime-Brain-Injuries.aspx.

9. "Army Finds Promise in Durable Material for Future Soldier Combat Helmets," Army Research Laboratory Public Affairs, October 11, 2017, https://www.army.mil/article /195105/army_finds_promise_in_durable_material_for_future_soldier_combat_helmets.

10. John Lloyd, "Impact Absorbing Composite Material," United States Patent Application Publication no. US2015/0246502, September 3, 2015.

11. Quotations here and in the next sentence are from Dennis, interview with Bauman and Rasor, June 29, 2009, and telephone interviews with Bauman and Rasor, 2017.

12. "Findings of Fraud and Other Irregularities Related to the Manufacture and Sale of Combat Helmets by the Federal Prison Industries and ArmorSource, LLC, to the Department of Defense," Investigative Summary, Office of the Inspector General, U.S. Department of Justice, August 2016, https://www.prisonlegalnews.org/news/publications/office-inspector-general -findings-fraud-and-other-irregularities-related-manufacture-and-sale-combat-helmets-2016/.

13. Kenner, telephone interviews with Bauman, 2017.

14. Chris Isleib, Office of Secretary of Defense, email response to Bruce Lambert, *New York Times*, regarding Lambert's questions provided by the Defense Logistics Agency, December 26, 2007.

15. Tammy Elshaug, interview with Bauman and Rasor, December 2, 2008.

16. Paul Scharre and Lauren Fish, "Reform Weapons Training to Protect US Troops from Brain Injury," The Hill, May 7, 2018, http://thehill.com/opinion/national-security/386468 -reform-weapons-training-to-protect-us-troops-from-brain-injury.

Index